Spectacles of Waste

Spectacles of Waste

WARWICK ANDERSON

polity

First published in 2024 by Polity Press

Polity Press
65 Bridge Street
Cambridge CB2 1UR, UK

Polity Press
111 River Street
Hoboken, NJ 07030, USA

ISBN-13: 978-1-5095-5740-0
ISBN-13: 978-1-5095-5741-7(pb)

A catalogue record for this book is available from the British Library.

Library of Congress Control Number: 2023944563

Typeset in 11 on 13 pt Sabon
by Cheshire Typesetting Ltd, Cuddington, Cheshire
Printed and bound in Great Britain by TJ Books Ltd, Padstow, Cornwall

The publisher has used its best endeavours to ensure that the URLs for external websites referred to in this book are correct and active at the time of going to press. However, the publisher has no responsibility for the websites and can make no guarantee that a site will remain live or that the content is or will remain appropriate.

Every effort has been made to trace all copyright holders, but if any have been overlooked the publisher will be pleased to include any necessary credits in any subsequent reprint or edition.

For further information on Polity, visit our website:
politybooks.com

Contents

Introduction

Modern Excrementalities and Postcolonic Biopolitics

My subject is shit, unavoidably. For most of us, it is a subject that elicits disgust and unease, fears of contamination and transgression, feelings of anxiety and insecurity. Shit is something we humans want to eliminate, remove from our vicinity, avoid touching or smelling – especially when it is others' shit. But the same intense disavowal and fervent distancing, the same desperate repression and displacement, also seem to compel us to analyze, calculate, survey, and write up the very thing that causes so much discomfort and distress. And so, that which ought to go unnamed in fact generates unceasing discussion in the humanities, the sciences, and popular culture. Our perpetual denial and rejection of shit force us to think again, and to think differently, about what it means to be human and modern; our reactions to shit make us speculate on how we might become more hygienic, secure, and civilized citizens. Surely, efforts to control shit, the seductive stool, are among the infant's first opportunities to theorize the world. Do we ever stop pondering shit in this way? In other words, it may be shit and its disavowal that make us "modern humans" – in as much as we aspire to that status or are permitted that status. The response to shit serves as the means to figure out who we are and who we want to be – and what we are not. Shit, then, can be good to think with and against and beside.

In this book, human shit proves to be a remarkably capacious subject, one that allows us to mess with wastewater surveillance or sewage snooping, the gut microbiome, the politics of *colon*-izing others and ourselves, perceptions of pollution and danger, feelings of revulsion and repugnance, along with an array of fecal fascinations and repressions. A dirty subject that touches on excremental colonialism, the fecal body politic, the latrinoscene, anal characters and anal pleasures, stool fetishes, intestinal intoxication, turd romancing, gut buddies, the helminth underground, kitsch tropical laboratories, and the rise of the biomedical sciences in the twentieth century. A subject, as we shall see, that ranges from environmental epidemiology and contemporary genomics to literary figuring of the grotesque or carnivalesque, to the reveries of psychoanalysis and social theory, and to the proliferation of shit art. Among this cornucopia of modern excrementality, my principal goal has been to reveal the scato-logics of the life sciences and population health during the past century and more; that is, to perform a procto-ontology of those sciences that help to make us "developed" and civilized.

Shit is a subject that has long preoccupied me – like everyone else, I expect. In the 1990s, I began writing articles with titles such as "Excremental Colonialism" and "Crap on the Map," observing pervasive and durable fixations on the intimate relations of bodily wastes and states of health or disease – offering critical historical studies tracking the cloacal captivations of modern public health. I was especially interested in how toilet practices became technologies of "whiteness," and how other races became "wasted" and pathologized under the sign of shit. By 2020, however, scarcely a whiff of this earlier research lingered, my thoughts about the history of shit having been, so it seemed, thoroughly evacuated. Then came

COVID-19 and a renewed sense of excretory urgency – or at least, a renewed apprehension of the need for further critical inquiry into the paradox of how shit, repellent yet also alluring, can both devalue and acquire value.

Soon after the pandemic began, I found myself engrossed in discussion of the merits of sewage surveillance, as one does. I was in Hobart, Tasmania, talking with the state director of public health, whom I have known since we were medical students, about the growing political and popular obsession with wastewater testing for the novel coronavirus, SARS-CoV-2. At the time, we faced the conundrum that any fecal findings would be inconsequential without additional public health intelligence. In communities with adequate individual testing and contact tracing, like Australia, the results of wastewater epidemiology did not matter much; certainly they were rarely sufficient in themselves to determine health policy. Yet everyone then was clamoring for more sewage studies, fetishizing excrement, demanding their sentinel wastes be registered and archived. They were stuck on the promissory value – and potential risk – of shit. The virus was reschooling us in stools. As Emma Garnett and colleagues note in *Critical Public Health*, "waste has come to matter in distinct ways during the COVID-19 pandemic, providing opportunities to re-think waste as a problem."[1] And in compelling us thus to rethink waste as a problem, the pandemic inevitably insists on us interrogating our sense of self, prompting us to reassess our body's relations with itself and the world. Feminist theorist Judith Butler observes:

> The definitive boundaries of the body pressured by most forms of individualism have been called into question as the invariable porosity of the body – its openings, its mucosal linings, its windpipes – all become salient matters of life and death . . . How, then, do we rethink bodily relations of interdependency, intertwinement, and porosity during these times?[2]

Or, just as urgently, how do we recognize continuing scatological resistances to, and denials of, any new-found entanglement and porosity? In this sense, every scatology is fundamentally an eschatology too, a theory of the destiny of humanity.

Meanwhile, the planetary excremental burden continues to surge. Along with our domesticated animals, we produce more than four trillion kilograms of fecal "biomass" each year. Not surprisingly, China and India together contribute a quarter of the world's human shit, with Brazil and the United States following close behind.[3] Much as we want to deny it, we are wallowing in the stuff. A lot of the non-human animal manure ends up as fertilizer in agriculture; some human waste in developed urban centers is removed, flushed away through sewage systems, then treated and destroyed; some in poorer spots is dissolved in septic tanks; and most turds in the remainder of the world just get deposited in the surrounding environment. Careless disposal of shit and failures of sanitation undoubtedly promote the spread of enteric pathogens, which can cause cholera, typhoid fever, poliomyelitis, and a range of diarrheal diseases.[4] There is no gainsaying the risks of indiscriminate contact with feces, of inept and discriminatory fecal management, especially affecting the poor, marginalized, and dispossessed. And yet, most epidemiological studies and molecular research concerning human stool in the past thirty years have concentrated on middle-class communities and sewersheds, on the wealthy sewered 30 percent, where such diseases are rare. Predominantly, the biomedical commitment has been to make this "developed" excreta data rich and informationally loaded. The vast enterprise of datafying and digitizing stool, of abstracting and securitizing feces, occurs disproportionately where the threat is least. In sewered parts of the world, where most wastewater epidemiology and gut microbiome research take place, shit is rarely infectious and noxious. Accordingly, the datasets and other symbolic equivalents derived from it, so far, have demonstrated little utility in public health and clinical practice. There must be some other explanation, then, for these relentless and extravagant efforts to abstract and to objectify and to securitize our beguiling yet still fearsome shit.

Why shit? Why now? These are questions we ask ourselves almost every day, sometimes insistently, with deep-seated feelings of necessity. Despite such universal physiological needs, the metaphysical expression of these irrepressible urges has not commonly been subject to critical scrutiny. What I want to do here, however, is look closely at how defecation and intestinal stasis – a kind of anal dialectic of flow and blockage, perhaps the most visceral dialectic of all – have prompted and shaped knowledge practices in epidemiology, microbiomics, public health, psychoanalysis, social theory, literature, and art. I am interested in what shit comes to symbolize in these domains. As we shall see, each study of defecatory possibilities and excremental imaginings inevitably leads to another, then another. Time and time again, shit conjures a sense of danger and disorder, a perception of open orifices and permeable boundaries, an incessant threat to corporeal integrity and safety. The disciplines of the humanities and biomedical sciences generate and regenerate possible solutions to this most proximate problem of disorder.[5] Thus, in opposition to shit's defacements and defilements, the arts and sciences together serve to provide alternative models of the secure citizen, the private civilized person, as well as to constitute ideal hygienic polities – modern collectives of mutually untouchable, or immunitary, individuals. The disciplines come to anathematize and mobilize shit, using it to police spatial and political boundaries and to ensure, as historian Dominique Laporte put it, the "sphincteral training of the social body."[6] But then, implacably, shit returns to stain such carefully laundered whites – a seemingly indelible brown presence.

I want to explore in these pages – through a veritable "history from below"[7] – what it is that drives us to know our shit. As matter out of place, excrement disturbs our sense of order and identity, breaching boundaries, disrupting classificatory grids, declassing and defiling that which we want kept pure and secured. It shows, ironically, just how hard it is for modern humans to live with our bodily functions. We respond to excremental danger either through expulsion and elimination, flushing shit away, wiping it off, banishing it to the sewer, or through sublimating whatever bits hang around, through abstracting this base matter, objectifying it in ways that might render it just another dataset or collection of images. Thus, wastewater epidemiology and molecular analysis of the human gut microbiome – along with much social theory, psychoanalysis, philosophy, anthropology, literature, and art – may be regarded as apotropaic genres, modes of warding off the perceived excremental threat, incantations against the supposed troublemaking powers of excrement. We are driven to inscribe, calculate, and digitize our shit, to classify and organize our poo, to reimagine the colon as sanitary, even Elysian. In the form of symbolic equivalents, as an array of data substitutions, shit looks neat and tidy, serially ordered, even informative; certainly, it no longer smells. In classifying shit, we re-class it, trying to exclude whatever makes it nasty shit.

The parts of this compelling redemptive process, this rarefication, that interest me most occur within what may be called the "ritual frame" of the modern laboratory, a sanitary place, delibidinized, where shit can be conjured as just another specimen or sample, transmuted into a surrogate object, reduced to a controllable assemblage of cells and molecules, an aggregation that can be cautiously manipulated. Accordingly, we try to recuperate stool as a kind of fetish, an autonomous power-object that represents and disguises the social and political work that its abstraction and refinement may be doing. Nonetheless, this alienation of excretions – what Friedrich Nietzsche called "the pathos of distance"[8] – proves unsustainable. Despite desperate attempts at sublimation, despite our multiple voidances, shit keeps coming back to bite us in the bum. While we try to transform it into a safe object – displacing it, inscribing it – shit remains "abject," something we thrust aside or alienate but from which we never really part.

It is telling that disquisitions on shit so often get simplified into structural binaries, into simplistic contrary ontologies: hygienic or filthy, pure or dangerous, civilized or natural, modern or primitive, white or brown – with the first term of each juxtaposition privileged. We seem to lack the ability to reason ecologically or even simply to think symmetrically, to think with balance, when it comes to shit. In this dialectic of the sphincter, the negative or inferior polarity is violently projected onto other races, characterizing them as inherently excremental and polluting – dangerous, even terrorizing, types. I have called this the *colon*-izing process, meaning the inscription and stereotyping of shit in order to stigmatize or subjugate others, especially other races – indeed, it is an essential part of the constituting of "other races." It involves relentless disparagement and degradation, signifying specific types of people as corrupted and defiling, as promiscuous defecators, dangerous shit spreaders. My goals here are to mark this production of difference, to show its undecidability or instability, and to explore critically the slippery dialectical interplay. The attribution of excrementality, or excretory difference, is perhaps the most intimate and powerful impression that colonial projects can make on subject bodies. Of course, by "colonial projects" I mean not only formal imperial regimes, but any rationalized power imbalance or political asymmetry. Empire offers models for such colon-ization, but it scarcely exhausts the practice. This process explains my recourse to the queer appellation, "*postcolonic* biopolitics," where the "post" in postcolonic indicates anti-colon-ial critique, a stepping aside from durable colon-ic thought styles, rather than their easy and uncomplicated elimination or succession.

And yet, as I hope will become clear, even ostensibly civilized scientists and epidemiologists, even the most white and hygienic of them, cannot hide forever their dirty secrets. Inevitably, they come to realize their own internal colonization, a reluctant yet undeniable creeping self-incrimination. Their absolute fecal discriminations and typologies thus fall apart and decompose. Hence the perceived need for recurrent datafying and elevating of their own stools, a sort of eternal reiteration of the civilizing process, purification rituals that sometimes provide additional benefits of producing intelligence to police excremental boundaries and ensure defecatory compliance. Philosopher Bruno Latour pointed out we have never been modern[9] – implying we can never be completely unencumbered by shit; it will always be around for us to try desperately to purify and sublimate.

What I want to do is scrutinize critically the infrastructures that make it possible for us to imagine ourselves as modern humans – to make ourselves up, or be made up, as modern persons. By "infrastructure" I mean the range of discourses and installations that structure society's waste disposal systems. Such scrutiny involves reckoning with the development of multiple classification systems and modes of standardization that work in various social worlds, retaining significance as they cross over boundaries.[10] It requires us to look at how this networked interoperability of our systems of thought can construct realities and constitute identities and polities – how it can "set parameters within which social action takes place," validating one set of social practices above others, placing ontological guideposts.[11] I am referring also, of course, to what Michel Foucault described as the *dispositif* or biopolitical apparatus, the heterogeneous and patchy ensemble of institutional, administrative, and discursive mechanisms that enhance and maintain the exercise of power relations.[12] Thus, the excremental *dispositif* includes sewage systems, flush toilets, public latrines, scientific laboratories, epidemiology, microbiomics, public health regulations, datasets, and digital files; it incorporates psychology and social theory, literature and art, language itself. As Giorgio Agamben put it, such biopolitical apparatuses have "in some way the capacity to capture, orient, determine, intercept, model, control or secure the gestures, behaviors, opinions, or discourses of living beings."[13] Such a *dispositif* or infrastructure, according to Foucault, is a "formation which has as its major function at a given historical moment that of responding to an *urgent need*."[14] That is, generating a *dispositif* is a bit like the act of defecation.

It is worth pointing out what this history of our excremental present does not do. It hardly touches on other human excretions and seepages, whether urine, semen, menstrual blood, mucus (or phlegm and snot), sweat, tears, and so on – though inquiries into these additional challenges to personal identity and bodily sovereignty would surely be illuminating. Nor do I consider non-human animal wastes, the piles of shit produced by livestock and domestic pets, since their shit seems to display distanced relations, if any, to our own human bodies, with therefore less impact on our sense of ourselves. Additionally, I have rejected here our many discards, such as ordinary garbage, leftover food, plastics, toxic chemicals, spent nuclear fuel rods, personal protective equipment (PPE), other medical waste, and so on, and on – since such detritus appears to me to imply connections with individual bodies and bodies politic quite distinct from intimate and privy responses to human feces. The closeness of bodily attachment to common physical trash is comparatively negligible; the antithetic corporeal symbolism is absent or paltry; and the ease of alienation is considerably greater. At the same time, my analysis does draw on several pioneering studies of trash or garbage or rubbish and on the recent profusion of "discard studies."[15] In particular, my exploration of postcolonic biopolitics seems to confirm general arguments in discard studies for the importance of the analytic defamiliarizing of waste; observing the vacillations in value of waste; recognizing the classification of waste as an operation of power; and accepting the need to specify and situate any delving into heterogeneous "waste." It behooves us to remember that things and people can be wasted in diverse ways and to different ends.[16] In writing what is, at least implicitly, an ecological critique of costive excremental humanism, I am not trying to assert any universal theory of waste – but nor would I hastily deny the pertinence of my findings to other environmental domains.

I hope to show here that shit has become one of the great spectacles of modern life, perhaps the most pervasive and compelling spectacle of all. Though we continue to insist it is nowhere, we realize deep down that it is everywhere. Through this exercise in postcolonic biopolitics I seek to make visible again the shit we fervently deny, to render it tangible as an actor in our lives, to reveal its political doings, to suggest how we might think it otherwise, ecologically. That is, I hope to squeeze out its vast and transgressive and complex messiness into a small book.[17]

1

The Sewage Panopticon

"Waste is a religious thing," muses Don DeLillo in *Underworld*. "We entomb contaminated waste with a sense of reverence and dread. It is necessary to respect what we discard."[1] In the novel, DeLillo has his principal character, Nick Shay, an Arizona garbage management executive, attend a conference in the Mojave Desert in 1978 on the future of waste. During a run through the desert between sessions, Big Sims, an African American delegate, reflects on his experiences of waste, telling Nick, "The thing about raw sewage . . . you treat it with loving care. You route it through bar screens way underground. And pump it up to settling tanks and aeration tanks. And you separate it and skim it and nurse it with bacteria." All waste, Big Sims assures his colleague, defers to shit and its caregiving. This careful wasting process has broader implications. As Jesse Detwiler, a UCLA waste theorist, tells them at the meeting, civilization arose out of the perceived need to discard waste, to set it aside and reprocess it, always with a sense of awe and trepidation. Disposal of shit, the waste theorist asserts, "forced us to develop the logic and rigor that would lead to systematic investigations of reality, to science, art, music, and mathematics."[2]

Detwiler might have added epidemiology to his list. The development of interest in the microbiology of feces paralleled the rollout of sewage systems in urban centers during the late nineteenth century. Sewers offered enticing new vistas for bacteriologists and epidemiologists studying patterns of infectious disease. In freshly sewered cities like London and Boston, a generation of microbe hunters flocked to the expanding repositories of wastewater, hoping to cultivate novel germs, to find species native to feces, and so make a name for themselves. These public health experts would spend a lot of time on their knees collecting sewage samples from local manholes, then analyze the bacterial composition of the waste back in their laboratories. Of all the materials that might reveal microbial secrets, feces appeared to be among the more hazardous, yet also the more compelling, possessing a special epistemic allure if not authority. Even so, most of these pioneering excrement sleuths eventually gave up in disappointment, sometimes ruing their initial enthusiasm as a passing folly. The promise of large-scale wastewater surveillance, the value of fecal intelligence, would not be fully realized until a century later. And then, later still, with the emergence of the COVID-19 pandemic, wastewater epidemiology suddenly went viral.[3]

The exuberant discovery during the COVID-19 pandemic of the full informatic potential of human waste coincided with growing concerns about the reliability and safety of aging urban sewer systems. Since World War II, the expanding cities of Western Europe and North America had put greater stress on their decrepit infrastructures, leading to fears of sewage seepage and spillover. Abatement of water pollution became prominent on the agenda of emerging environmental movements. In the past few years, this sense of a sewage infrastructure crisis has heightened, in part because of the likely overload from storms or "large rain events" and flooding, consequences of human-induced global heating.[4] "Fecal matter represents a clear and present danger to that most precious right of access to clean water, contaminating natural sources like streams and bubbling up into bathtubs and sinks due to faulty septic infrastructures," a journalist told *Guardian* readers in 2022.[5] Reports of possible cyber sabotage of American sewage systems also float around the internet.[6] Even wealthy white sewered communities have started to worry about the prospects of contamination with human waste, the return of the expelled, the spillage of discards, the imminent deluge of sludge, the breakdown of barriers between hygienic bodies and their excremental products. Such anxieties about the permeability of boundaries that once seemed so robust – fears of what has been called the "fecal peril"[7] – reinforce the need to comprehend and contain our supposedly dangerous wastes, and thus to engage in wastewater epidemiology. When you cannot safely eliminate shit, you can try to subjugate it by objectifying it – you make it calculable and abstract, neutralized in biostatistics.

It is tempting, however, to reduce this story to a homily about technical advances and improved medical utility. Perhaps biomedical scientists simply got better at revealing the microbial mysteries of ordure. They learned how to factor shit into the epidemiological calculus. There is no denying the appeal of ever more information about the prevalence of potentially pathogenic microorganisms in human populations, even if data are often redundant or belated and the health implications obscure. The medical rationale cannot be dismissed. What I find more interesting is how fascination with human excrement so often is out of proportion to its likely utility, surpassing any use value. Even when feces fail to disclose much of practical benefit, we seem driven to delve ever deeper, searching for the microbial secrets that must surely be concealed in them. This is no relaxed quest for knowledge, no naïve following of the science – something more heartfelt is at stake. In *The Varieties of Religious Experience*, philosopher William James warned against facile recourse to "medical materialism," the supposition that scientific explanation and clinical functionality will always exhaust or substitute for symbolic import and spiritual meaning.[8] In this chapter I want instead to explore the ways in which language, culture, and social practice have shaped our apprehensions of the epidemiological value of human waste, thus resisting any easy, or at least conventional, philosophical refusal or sociological denial. That is, I seek to understand some epidemiologists' "obsessive preoccupation with the visceral and excrementitious subject," as Aldous Huxley once called it.[9]

COVID Down the Toilet

Although fecal transmission of the novel coronavirus, discovered in the Chinese city of Wuhan at the start of 2020, was rapidly discounted, the pressure to find fragments of it excreted in sewage proved hard to resist. Before long, experts in wastewater epidemiology were scouring sewers for molecular markers of the emergent virus, filtering and processing the discharge for pieces of its nucleic acid, thereby revealing evidence of outbreaks in the catchment areas or "sewersheds." Inevitably, scouting for fecal intelligence could occur most readily in places that are securely sewered and possess adequate laboratory and public health infrastructures. (In the United States, however, only about 80 percent of dwellings are sewered, while the rest rely on septic tanks.) Soon, though, public health officers in even the more lightly sewered parts of the world were rummaging through available excreta to find early warning signals or repeated confirmations of viral spread.

At the end of February 2020, Dutch researchers identified "probably unviable" SARS-CoV-2 in effluent from Schiphol Airport. The next month in Massachusetts, scientists found copious traces of virus in diluted and mashed up fecal specimens, revealing its "penetrance" in "specific vulnerable

communities." Across six days in April, Australian investigators detected two positive "grab samples" in a local sewage
treatment plant. In June, even India could report high "genetic
loading" of fecal material in its meager sewage infrastructure,
declaring the "bottom line" was that wastewater surveillance
provided a reliable "early warning signal" for COVID-19 – as if
any further evidence of viral prevalence was needed there. By
the middle of 2020, there were pleas not to leave Africa, with
its minimal sewerage coverage, invisible in the fecal panopti-
con. No human waste left unsampled became the mantra, if
not the practice.[10]

Early in April 2020, *Nature* confirmed that "Routine
wastewater surveillance could be used as a non-invasive
early-warning tool to alert communities to new COVID-
19 infections."[11] The prospect of fecal detection of the virus
excited strongest interest where the pandemic was escalating
in advance of adequate individual testing procedures. In effectively revealing our "dirty secrets," wastewater epidemiology
appeared to offer a convenient alternative to failed contact
tracing methods.[12] Sampling compliant bodily discards could
substitute for swabbing refractory bodies. A Dutch microbiologist noted: "Not everyone is getting tested, but everyone
goes to the bathroom. It's nice to have an objective tool that
isn't dependent on willingness to get tested."[13] The initial
reports bruited waste's sensitivity as an anonymous "early
warning tool."[14] After all, they insisted, excrement contained
the indelible "signature" of SARS-CoV-2. An Indian journalist
observed: "Scientists around the world are peeping into poop
and wastewater for the novel coronavirus to come up with
surveillance techniques." Wastewater sampling constitutes an
appealing "non-invasive early-warning tool" since it obscures
individuals and focuses instead on the sewer as a "mirror of the
population."[15] Accordingly, any blame for harboring the virus
might be deflected from specific persons onto whole communities and territories, or, more commonly, onto the supposedly

errant and otherwise stigmatized groups within them. Those who find delivering personally a fecal specimen shaming turn out to be comfortable and relaxed when their evacuations are distant and unnamed, rendered uniform, safely distributed, unacknowledged, alienated.

While the published results of wastewater epidemiology – its multiple polished inscriptions – attract most attention, the process of rendering effluent medically significant, of making shit signify, depends on a network of invisible technicians, ranging from dutiful defecators to waste samplers, to laboratory assistants, to microbiologists, to molecular biologists, to public health statisticians. It takes a lot of effort and negotiation to stabilize these connections, to articulate between skill sets, in order to produce reliable results. Most parts of the world lack the infrastructures and technological norms necessary for effective alignment of the components of wastewater epidemiology. Without sewage systems and conventional flush toilets, there is no "catchment," and it is impossible to render human excrement *epidemiologically* legible – even if individual specimens might still acquire medical meaning. Where sewers are more-or-less functional, taking representative samples from them upstream of chemical treatment demands considerable prowess and artistry. This is not simply a smelly "bullshit job."[16] Frontline samplers must develop a sense of a "good" portion; they learn how to isolate information-rich effluent, to contain and preserve it, to transport it without contamination. In the laboratory, the sample gains further value through the activities and interactions of scientists and technicians, who force its sloppy constituents through various preparations and procedures, including culturing, biochemical testing, centrifugation, polymerase chain reaction (PCR) amplification, reverse transcription, genetic sequencing, and so on. Then, those accredited as senior scientists exercise their expert judgment, arraying data in tables and subjecting them to statistical tests. At the end of this laborious chain, the results are *written up* as "wastewater epidemiology," a compelling sublimation of potentially pathogenic specimens, thus authorizing shit as a signature.

What just happened there? I have sketched how human refuse is processed and mobilized into the modern scriptural economy or, rather, elevated onto a global digital platform.[17] It is refined into what used to be called paperwork but might now be designated digital code – turned, in other words, into big data. One may thus imagine wastewater epidemiology as a vast excremental writing machine. Late last century, Bruno Latour urged us to look more closely at how "paperwork," whether files or dossiers or reports, can make scientific facts visible and then mobilize them around the world. He wanted us to see how facts are registered and defined, dressed up as immutable or at least stable entities, contrived to be combinable and circulatable. "Mobilization is not restricted to paper," he wrote in that pre-digital era, "but paper always appears at the end when the scale of this mobilization is to be increased."[18] As scientists accumulate "traces" and simplified inscriptions, they produce harder facts. Once abstracted, these resources may be further mobilized, or scaled up and down, through more "paperwork," to use Latour's metaphor. Therefore "by working on papers alone, on fragile inscriptions which are immensely less than the things from which they are extracted, it is still possible to dominate all things, and all people."[19] The reference to paper may now seem quaint, but Latour's argument still pertains to the contemporary epidemiological digitization of waste. It confirms the observation of DeLillo's imaginary waste theorist, Detwiler, that enacting the alienation of shit necessitates "the logic and rigor that would lead to systematic investigations of reality" – to more toilet paperwork, in other words.[20]

Initially, the collection, registration, and coordination of data from multiple sources represented a formidable bureaucratic challenge. In the United States, for example, thousands of sewage works and public health laboratories were accumulating and analyzing millions of wastewater samples by the end of 2020, hoping to find viral fragments in the fecal deluge. In Massachusetts, Biobot, a wastewater analysis start-up – which had begun by gathering specimens in 2017 from a manhole in Central Square, Boston, looking for traces of illicit drugs – expanded to trawl for evidence of SARS-CoV-2. Its surveillance efforts soon rivaled the piecemeal and scattered contrivances of government bureaucracies. Before long, more than 170 wastewater facilities in thirty-seven states regularly sent fecal samples to Boston for Biobot's expert scrutiny.[21] At first, most wastewater surveillance was intended to warn of COVID-19 outbreaks in areas resistant to individual testing – and later, to substitute for PCR and rapid antigen testing where they were wound back. As the coronavirus continued to evolve, attention turned to detection of possible new variants, otherwise unknown mutations. As Mariana Matus, the MIT computational biologist who co-founded Biobot, stated: the "holy grail would be able to identify completely new threats that nobody knows about yet before we know them."[22] Such strange RNA sequences or "cryptic lineages" promised moments of excitement among the excremental drudgery in the laboratory, but they rarely influenced any specific public responses, not until confirmed through conventional clinical testing and contact tracing. Often it was discovered that the supposed new variants came from contamination with rodent or other animal feces, not from human discards.[23] To coordinate the inundation of wastewater information, in September 2020 the Centers for Disease Control and Prevention (CDC) launched the National Wastewater Surveillance System, to assemble all American information on viral shedding in human feces. But it took another eighteen months for its waste dashboard – the poop data tracker – to go public, revealing sewer contents from thirty-one states. At Biobot, Matus told the *New York Times* that the government should have relied instead on the private sector to track the nation's sewage.[24]

In April 2022, some two years into the pandemic, Willemijn Lodder, a Dutch researcher who had been among the first to detect coronavirus fragments in human waste, proudly showed an inquiring journalist the contents of her laboratory refrigerator. Lodder had assembled thousands of plastic bottles containing cloudy brownish liquid. Wearing gloves, she daintily took a jar from the shelves, shook it a little, then gazed reverently at its contents, without opening the lid. "This is it," Lodder told reporter Maarten Keulemans, "Half a liter per jar. Straight from the sewer. Every morning the courier brings them here." In the past two years, wastewater epidemiology had proliferated around the sewered world, following the Dutch lead. Even though the incidence of SARS-CoV-2 was then declining in the Netherlands, couriers continued to bring 1,200 jars of sewage each week to the cold storage at Bilthoven, four samples from each of the country's 300 waste treatment plants. As Utrecht molecular biologist Ana Maria de Roda Husman put it: "A next variant could lead to more recordings and more distribution. So you want to keep a finger on the pulse [*een vinger aan de pols*]."[25] Sewage distillation, surveillance, and registration evidently are here to stay.

As the pandemic evolved, it rapidly became clear that positive sewage results could indicate the presence "in real time" of the novel coronavirus in the catchment area, which might encompass the nation or be confined to a single building or institution or ship – in practice, surveillance often concentrated on quarantine hotels, hospitals, prisons, aged care facilities, university dormitories, schools, and, increasingly, airplanes.[26] Then again, detection of viral fragments may merely signify a history of infection there or that someone had shed the virus when passing through. Not much is known about variability in fecal shedding of coronavirus; nor can we assess the confounding effects of different kinds of sewage systems, what has been called "the complexity of the wastewater matrix."[27] Wastewater epidemiology offers enigmatic warnings even as it provides no detailed information about current transmission patterns. According to de Roda Husman, interpreting raw sewage is a bit like "aura reading."[28] Testing nasal and throat swabs and contact tracing, where feasible, give more granular and definitive accounts of viral spread. It remains hard to infer anything useful from detection of SARS-CoV-2 in aggregated sewage, to distill any guidance for a targeted public health response. The relevance of picking up the novel coronavirus in sewage diminishes even further when most of the population is at least partly immunized against it. And yet, proposals to render shit epidemiologically legible have continued to exert a symbolic compulsion out of proportion to their limited utility in public health management.

The troubled quest for ever more *precise* wastewater epidemiology exposes current deficiencies in mapping disease distribution across large, homogenized, and anonymous populations. Precision public health seeks to address this old problem of how to relate population data to individual cases or to specific persons who need to be targeted. No surprise, then, that in response to these difficulties exponents of more precise "digital medicine" have begun calling for a "smart toilet platform" that would altogether erase the sewage system from the epidemiological calculus. Using a "passive" and automated approach, the smart toilet would allow consenting defecators simply to poo into a "mountable bidet-style attachment equipped with modules for automated faecal sample collection and processing [and] faecal RNA isolation and detection with in-situ ultrafast nucleic acid amplification tests."[29] The proposed smart toilet could detect "defecation events" and automatically scan for SARS-CoV-2 and other viruses before they are flushed away, linking the results to Bluetooth contact tracing systems. Ideally, it would take only fifteen minutes from defecation to notification. Fortunately, no one need handle their own stools to obtain specimens. Bringing the toilet into the "internet of things" therefore would greatly enhance population surveillance, though it would do so at the cost of anonymity and privacy. It would permit long-term testing for the coronavirus as it becomes less novel – as it is normalized as endemic.[30] Moreover, the smart toilet enables further wide-ranging and intimate profiling of willing and dedicated defecators. Sensors might also detect body temperature and oxygen saturation; stool morphology analysis can identify diarrhea; while drugs and other disease organisms are readily tracked. Promoted as a contribution to "citizen science," this individualized fecal surveillance mechanism promises to re-write our personal bio-narrative, fashioning new excremental abstractions of the self.

Sentinel Feces?

Long an object of medical fascination, human excrement occupied a key position in nineteenth-century theories of disease transmission.[31] Studies of the spread of cholera and typhoid, among other "enteric" diseases, focused the attention of sanitarians on the dangers of feces, underscoring the need to monitor and contain polluting human wastes. Statistical analysis of cholera outbreaks in large cities such as London vividly revealed the dangers of contamination of water and food with sewage, rationalizing an aversion otherwise based on deep-seated distaste.[32] At first, most sanitarians concentrated on the chemical or non-vital emanations of shit, its putrid and corrupting attributes, manifesting mysteriously either in noxious smells or some foul poison or toxin. Accordingly, they warned the public to avoid defilement with dejecta, to isolate their base matter and send it as far away as possible. Although produced by humans in the most intimate of circumstances, shit appeared to represent the most fearsome and public of *environmental* threats, prompting civic debate on how best to ameliorate its disposal and elimination, leading to the laying down of urban sewage systems and the invention of flush toilets.[33] As germ theories became ascendent late in the century, bacteriologists made visible a horde of miniature life forms – bacteria, vibrios, micrococci, and parasites – in human stools, some readily associated with specific infectious diseases, others apparently harmless. According to a Toronto, Canada, public health officer: "A large proportion of the disease to which the highly organized yet frail units of humanity are almost universally subject is caused by the waste products of human life."[34] This sudden materializing of disease agents reinforced longstanding worries about the pathogenic properties of human waste, boosting the urge to treat shit with care, even when fears of contact with feces still might draw more deeply on moral intuition than epidemiological reason.

The growing rationalization of the risks of human waste in nineteenth-century Europe and North America fostered the development of urban sewer systems that separate out and deflect bodily discards, whether typecast as environmental pollutants or as media hosting recently discovered pathogens. Fearful of the dangers of unregulated excrement, sanitarian Edwin Chadwick campaigned hard for the sewering of London in the 1850s, hoping to divert the contents of the capital's privies and cesspools onto surrounding farmland. When Joseph William Bazelgette eventually completed work on the city's model sewer system in the 1860s, he found it expedient instead to channel the torrent of sewage into the Thames downstream. Despite this controversial re-routing, Bazelgette's basic design soon was copied in the burgeoning urban centers of the "civilized" world, such as Paris, Chicago, Brooklyn, and Boston, among countless others.[35] In the 1860s, German pathologist Rudolf Virchow – a founder of social medicine, which recognizes economic determinants of health, and skeptical of bacteriology – organized Berlin's sewage system, hoping to remove baneful substances from the vicinity of the city.[36]

Sewer systems came to function as the excretory organs – the bowels – of modern urban systems.

"Toilets in modern water closets rise up from the floor like water lilies," wrote Milan Kundera in *The Unbearable Lightness of Being*. "The architect does all he can to make the body forget how paltry it is, and to make man ignore what happens to his intestinal wastes after the water from the tank flushes them down the drain." Kundera saw the expulsion of shit as the evasion of an onerous theological problem, the denial of responsibility for pollution, the refusal of the defects of creation. The technological advances of the nineteenth century – principally sewer systems and flush toilets – mean we remain "happily ignorant of the invisible Venice of shit underlying our bathrooms, bedrooms, dance halls, and parliaments."[37]

Initially, sewer engineers relied on simple dilution with water to dissolve and distance any fecal dangers. Toward the end of the century, especially in North America, most cities turned to novel techniques of sedimentation or filtration and chemical treatment to render effluent innocuous. Later, sewer managers learned to add selected "good" or "probiotic" microbes that could break down or decompose solid organic matter, processes enhanced by "activated sludge" intercessions in the 1920s and 1930s.[38] Thus, the sterilizing of shit went from being a chemical problem to predominantly a bacteriological field of inquiry. Its decay and dissolution increasingly depended on microbiological supplementation; while assessment of its safety after treatment required bacteriological testing, not chemical analysis. Measurement of microbial flora in treated sewage became a routine index of the effectiveness of purification technologies – a sign of any potential danger.

It should come as no surprise that the new generation of bacteriologists was fascinated by human waste and in thrall to its epistemic allure. A special sense of urgency and strain was evident in their labors. During the winter of 1902–3, C.-E.A. Winslow and D.M. Belcher from the "Sewage Experiment

Station" of MIT were collecting raw fecal samples through a manhole on the Dartmouth St. side of Copley Square, Boston. The future leaders of public health in the United States then cultured the diluted excrement on agar plates, discerning a wide variety of different organisms, which they classified in a Borgesian bestiary seemingly unrelated to our contemporary bacterial taxonomies. One elaborate microbial type appeared after another: spirillum, fluorescens, chromogenic, vulgaris, cloacae, liquidis, superficialis, subtilis, rhinoscleromatis, and ubiquitous. The valiant stool sleuths were sure that sewage represents an especially rich nutritive medium for germs, though they discovered no evidence of "true sewage bacteria," microbial flora unique to that environment. The scientists expressed relief in finding that white residents of the Back Bay appeared to defecate fewer "colon bacilli" than previously reported in the same amount of filthy London excrement, perhaps testament to New World virtue.[39] Early issues of New England's *Journal of Infectious Diseases* are replete with similar inquiries into the bacterial composition of sewage and discussion of how best to handle it and preserve it without exposing the investigator to grave danger.[40] But the connections of microbial flora in untreated sewage to the distribution of infectious disease among the sewershed's human population remained a puzzle. Rampant bacteriomania initially offered few answers. The reliability of raw feces as sentinel of infectious disease, as an index of public health, baffled epidemiologists – it was as if so much sanitary effort had gone into making shit invisible and insipid, nothing they could do would make it perceptible and purposeful again.

Sewage Information Mining

Despite the frustrations of trying to make excrement epidemiologically legible, sewage "information mining" did slowly become an established specialty.[41] Yet it has remained marginal and seldom figures in histories of epidemiology: it was, perhaps, the secret disgrace of the broader field. Nonetheless, those who continued valiantly to seek pathogens and biomarkers in raw sewage could boast a few successes by the 1960s. Through calculating the concentration of microorganisms or offending substances and the volume of waste in relation to the population of the catchment, they learned how to arrive at a rough approximation of the prevalence of microbes or toxins or drugs in the sewershed.[42] In 1939, microbiologists had found fragments of poliomyelitis virus, which spreads through fecal contamination, in untreated sewage from three North American cities.[43] Scientists at Yale later determined concentrations of polio virus in a Connecticut wastewater treatment plant before, during, and after an immunization campaign, showing that attenuated vaccine strains had persisted in the population for about three months.[44] In 1977, infectious or "wild" polio virus was detected in Israeli sewage more than a week before any clinical evidence of the disease could be found.[45] At the turn of the century, the World Health Organization (WHO) issued guidelines for wastewater surveillance to detect communities that still may harbor polio despite the international eradication campaign.[46] During the past decade, numerous enteric microbes, including hepatitis A, salmonella, and norovirus, have turned up in European waste collections, indicating community spread of these diseases. Surveillance capacity widened so as to trace microorganisms resistant to common antibiotics, as well as fungi such as *Candida auris*. During the 2002–3 SARS outbreak, that antecedent coronavirus was detected in wastewater collected from several Chinese hospitals.[47]

Until the COVID-19 pandemic, the most significant achievement of wastewater epidemiology was the surveillance of urinary effluent for drug biomarkers – an exercise in chemical detection, not microbial culture. In 2001, Christian C. Daughton,

chief of environmental chemistry at the Environmental
Protection Agency in Las Vegas, Nevada, proposed a program
of "non-intrusive drug monitoring at sewage treatment facilities."[48] While he expressed concern that drugs and personal
care products might damage vulnerable aquatic ecosystems, his
main interest was assisting law enforcement officers to crack
down on "abuser" communities. Within a few years, a team of
researchers was able to identify large quantities of cocaine in
the River Po and a range of urban wastewater plants in northern
Italy.[49] Since then, surveillance for drugs and their metabolites
– especially for cocaine, methamphetamine, and opiates – has
proliferated around the world. The widespread monitoring of
urinary excretion offers advantages of being non-invasive and
anonymous, as well as cheaper, more reliable, and quicker than
conventional personal surveys of drug usage.

All the same, critical sociologists argue that "wastewater
analysis constitutes drug use as measurable, countable and
comparable and, in doing so, enacts a homogeneous drug using
population in a bounded geographical space."[50] It stigmatizes
groups of users as untrustworthy, uncooperative, and criminal;
it directs attention to suppressing drug use rather than harm
reduction. Science studies scholars Kari Lancaster and Tim
Rhodes caution that while "wastewater analysis can provide a
snapshot of drug concentration overall, the method is blind
to the complex social dynamics that shape drug harms and
the transmission of viruses." It might provide early warning of
viruses yet "it cannot account for dynamic population patterns
or the specific social and behavioural practices that give rise
to outbreak events."[51] However, in response to the COVID-19
pandemic, Daughton has been roused to demand "a national
strategic plan [in the US] for wastewater-based epidemiology," in the hope that "this tool become a workhorse in [virus]
investigation."[52]

The successes of drug monitoring in the twenty-first century, albeit limited, reoriented wastewater epidemiology away

from microbiology toward molecular analysis, usually conducted by biochemists and computational biologists. Effluent
became subject to new modes of datafication and new styles of
inspection, registration, and calculation.[53] The dominant paradigm of drug metabolite testing also directed attention from
concerns about fecal contamination and environmental taint
toward identification of aberrant lifestyles and social pathologies, even if representations of sociality and behavior remain
partial and impoverished. Before, if a microbiologist detected
polio virus in excrement, it signified a biosecurity issue. Now,
if a biochemist or data scientist observes a biomarker, it usually
connotes criminal activity or moral turpitude, some style of
transgressive behavior. These methods and goals have reshaped
wastewater epidemiology, allowing it to constitute populations
in different ways, sometimes as potentially harmful to others –
a disease hotspot – sometimes as self-harming and licentious.[54]
The new datafication of shit certainly was never intended to
work as a technology of care.

The fastidiously cleansed city-state of Singapore, a laboratory of hygienic modernity, presents an illuminating example of
evolving tastes in wastewater surveillance. In the first decades
of the twenty-first century, the city's drug control authorities
tried sampling sewage, especially the wastes from foreign workers' dormitories, but results merely confirmed the efficiency of
national suppression of any illicit activities. Not even foreign
feces could be fashioned into a smoking gun. By 2010, scientists
at Nanyang Technological University were looking instead for
fragments of enteric viruses, the causes of diarrhea, in "human
faecal matter" extracted from the tropical sewers. They found
an astounding array of viruses, possessing surprising genetic
diversity in contrast to the supposedly homogeneous human
population, which suggested widespread occult gastrointestinal disease in the retentive city-state.[55] A few years later,
scientists from the National University of Singapore were
combing the island's sewers for signs of residual antibiotics and

multidrug-resistant pathogens, describing how hospital excrement in the tropics constituted a "cocktail" of antimicrobial
resistance.[56] But before the advent of the novel coronavirus,
most wastewater surveillance in Singapore was concentrated
on detection of arthropod-borne viruses, or arboviruses, such
as dengue, yellow fever, and Zika. These conditions often can
be mild, and therefore may be missed in clinical surveillance.[57]
Singapore's long-term preoccupation with monitoring sewage
meant it was well prepared to examine human waste for traces
of SARS-CoV-2 early in 2020. The National Environmental
Agency (NEA) sprang into action, assiduously gathering feces
for inspection, especially the effluent from migrant workers,
hospital inpatients, and university students.[58] At the beginning
of the pandemic, reporter Matthew Mohan followed an NEA
team of sewer gumshoes, watching them prise open a manhole cover near a migrant workers' dormitory. According to
lead scientist Judith Wong, inspection of the foreigners' waste
would complement the regular swabbing of the inmates of
the dormitory, indicating "whether or not infection control
measures have been effectively carried out." Feces had become
a litmus test for compliance with hygiene standards. If any
transgression was detected, Wong assured Singaporeans, then
"it will prompt appropriate follow-up measures such as swab
testing of the population to allow us to identify the infected
migrant worker."[59]

Interoperable Effluent

In March 2022, US President Joseph R. Biden, Jr. unveiled his new
COVID-19 strategy, shifting focus from managing a "crisis" to
dealing with the "new normal," to coping with endemic disease.
A crucial component of the scheme was enhanced wastewater
detection of emergent variants of the persistent virus, expanding the CDC's National Wastewater Surveillance System or

"poop tracker."[60] The National COVID-19 Preparedness Plan aimed "to improve our data collection, sequencing, and wastewater surveillance capabilities," mostly through "strengthening data infrastructure and interoperability."[61] In the plan, base matter was objectified as informatics, effluent became input for computational biology. The euphemism "wastewater" substituted for feces or shit. Not even "sewage" could be mentioned. To assuage our anxieties, that which cannot be handled safely would be dexterously manipulated into datasets of promissory value.

As never before, we are driven to make the unnamable speak an abstract language, to render it calculable and interoperable, to trust in numbers, to "datafy" and thus securitize our unmentionables. The impulse behind the irresistible rise of wastewater epidemiology therefore seems far stronger, more deeply visceral, than any purported technical advantage. As historian Dominique Laporte observed: "Shit ceases to be shit once it has been collected and transmuted, and only exists in the form of symbolic equivalents."[62] Increasingly, many of us feel the need to molecularize and calculate human waste: through meticulous analysis and conscientious recordkeeping we hope to transform our discards into objects that can be scrutinized safely at a distance.[63] The hygienist "overcomes the most visceral repugnance, rolls up his sleeves, and takes on the cloaca," vowing to make the unutterable speak.[64] This relentless and compelling task suggests a certain kind of bourgeois subjectivity – once a cultivation of "whiteness" – that demands separation from stuff that threatens to disrupt identity, requires the elimination of wastes or at least the objectification and distancing into informatics of all that we find abhorrent. As philosopher Roland Barthes put it: "when written, shit has no odor [*écrite, la merde ne sent pas*]."[65]

This brief history of wastewater epidemiology shows that white people, especially white men from the northeastern United States, have displayed a penchant for writing up and adjudicating the contents of sewers. We all shit, but histori-cally only elite white people, mostly white men, got to assay it within the ritual frame of the laboratory. Their dedication to transforming human waste into measurements and figures was, until recently, unrivalled. Now, competitors also are building their own excremental platforms, taking to the sewage stage, across Asia and Latin America, and sometimes in Africa and the Pacific too. Often non-white and female, they share the same commitment to hygienic modernity, to the practi-cal elimination and symbolic sublimation of hazardous feces from "civilized" communities, the same dedication to sewers as technologies of excremental securitization.[66] Once safely abstracted and quantified, human waste can serve yet again to police the boundaries of bourgeois respectability, to make object lessons, to stigmatize those bodies not up to scratch – the dangerous and transgressive defecators among us, the marginalized, the foreign, the disabled, those deemed shiftless and untrustworthy.

From the nineteenth century, modern sewage systems have channeled cultural and social imagination. More than just infrastructural aids to somatic release and relief, sewers have served as metaphor and model for the safe disposal of human wastes and wasted humans. Thus, Aleksandr Solzhenitsyn described the Soviet Gulag as "our sewage disposal system." As dissidents and minorities massed in the labor camps, they "strained the murky, stinking pipes of our prison sewers to bursting." Between 1944 and 1946, the wartime government "dumped whole nations down the sewer pipes." There was a "hygienic purging" of dangerous intellectuals and heretics from purified and enlightened Soviet society. "Through the sewer pipes the flow pulsed," Solzhenitsyn wrote. "The waves flowed underground through the pipes; they provided sewage disposal for the life flowering on the surface."[67]

2

The Waste That Therefore I Am?

Something both seductive and disconcerting about shit has prompted Europeans to speculate on human identity and corporeal integrity since "classical" times, long before they became "Europeans." Indeed, performances of defecation, or rather, their disavowal, may have initiated philosophical discourse. The term "analysis" comes from ancient Greek, meaning the act of loosening or releasing from a ring, from some annular design, such as decoupling from a mooring, or perhaps extruding from an anus. Feces came from a Latin word for dregs; defecation implied the removal of such impurities; while the Romans knew excrement as whatever is sifted out or refined – as happens in good argument or debate. In a rather curious locution, classics scholar Dan-el Padilla Peralta observes of Homer: "To contend with waste – the excremental variety in particular – is to see the ideal plop to the ground and the beautiful funkified."[1] According to Padilla Peralta, Plato's Socrates was "the sanitizer of sanitizers," whose excremental scenes revealed "the forms of privilege and oppression that cluster around waste relief." In ancient Greece, "waste morphed into a medium with a message," contributing to the "longue durée semiotics of ordure," and the "turdification" of philosophical speculation in Europe.[2] Evidently, excrement has been good to think with – perhaps especially in private, as interior monologue.

Like most things we are not supposed to talk about in public, shit lurks close behind many of our thoughts, intruding disturbingly on our minds. Wastewater epidemiology is merely the most recent attempt to make sense of the quality of human waste, to assay its value and its risks. Sewage surveillance can therefore be seen as yet another somatization of social theory, albeit heavily fortified with numbers and calculations. As an effort to objectify and alienate base matter, it perversely contributes to the long history of subjectivity, or more precisely, to histories of the fashioning of subjectivity. In this chapter, I want to consider some of the antecedent apprehensions of philosophers, psychoanalysts, social theorists, and cultural anthropologists concerning human waste, showing how these savants have stayed with the troubles of matter out of place. Moreover, I would like to know whether philosopher Peter Sloterdijk could be right when he claims that modernity, or civilization, entails the sterilizing and deodorizing of "merdocratic" and "latrinocentric" spaces.[3] If Sloterdijk's metaphoric allusions to privatized and deodorized modernity have merit, then they may allow us to give embodied form and full excremental weight to Bruno Latour's assertion that we have never actually been modern, never effectively toileted – we can never sustain the purification rituals needed to separate culture from nature, civilization from shit. "No one has ever been modern," Latour insists. "Modernity has never begun."[4]

But I am getting ahead of myself. The purpose of this chapter is to illuminate the immense philosophical or analytic appeal of human waste during past millennia, the cogency of excreta in explaining how we become modern humans, or rather, come to imagine ourselves as such. "'Humanism,'" psychoanalytic historian Dominique Laporte tells us, "could be defined by its penchant for waste."[5] What we think of as "human," or perhaps as a specific kind of human, may be whatever is left over when we thrust aside or eliminate that which is perceived as waste. In his lecture on the "sick soul," William James, an exemplary New England puritan, described the threat that "matter out of place" presents to personal identity: thus, any challenge to our sense of self must be regarded as "an alien unreality, a waste element, to be sloughed off and negated." The modern human thus will be "marked by its deliverance from all contact with this diseased, inferior, and excrementitious stuff."[6] But this process of self-definition is always inept and unfulfilled since our discards retain the potential to return and harm us. The patrolling of orifices and boundaries can never make them secure enough for us to relax, even for a moment. Relentless scrutiny and objectification of our excreta, their demarcation, their sterilization and deodorization, their securitization at a distance, can never be sufficient to reassure us for long. Shit may be rejected and flushed, but we never properly part with it. No matter how well toilet trained, we are faced with what historian Alain Corbin has called "the implacable return of excrement."[7] Or, in other words, we may not be modern after all.

Shit Histories

According to several French historians, the analysis of shit gives a plausible genealogy for the constitution of bourgeois subjectivity. In *Histoire de la merde*, published in the 1970s, Laporte describes the increasing privatization of excrement since the sixteenth century, taking place initially in spreading, densely packed European cities. When the space of defecation was incorporated into the family home, reordering patterns of domestic intimacy, it also became more private, secreted in a water closet or privy. Human manure had long been spread over agricultural land so as to fertilize the soil and enrich produce, but modern urban environments appeared to require expulsion of wastes and cleansing and deodorizing. One might shit in private in the family home but then the smelly and dirty excretions must be removed, sent as far away as possible. In particular, a lower tolerance for seemingly noxious odors animated this urban "policing of waste and . . . privatization of excrement." "The apprenticeship of the sense of smell," writes Laporte, "directed all its efforts against the stercus." As time went on, the centralized state apparatus intervened more forcefully, attempting to channel, to purify, and to sterilize the shit of city dwellers. In Laporte's view, "the state is the sewer"; it emerged in parallel with the reticulation of human waste; its role became the "sphincteral training of the social body."[8]

A few years after Laporte's literary discharge, Corbin returned us to the site of expurgation, immersing us further in coprological themes. He traced the development of special sensitivity to excremental odors in late eighteenth-century France, the rising fear of the dangers of putrefying excreta. Smell alone was pathogenic, smell alone could kill. "The privatization of waste," Corbin assures us, "encouraged the containment of excremental odors within strictly limited places," in the hope of curbing deadly emanations. "Abolishing the promiscuity of [public] latrines, keeping doors closed, and installing blow-off pipes were indispensable preliminaries for that disciplined defecation deemed essential for the elimination of stenches."[9] The space of excretion was not the sole locus of dangerous odor. The fetidity of the laboring classes excited similar caution, spawning the same distrust of the feculent smell of the poor. The bourgeoisie felt that "making the proletariat odorless would promote discipline and work among them." It might, in well-ventilated spaces, make the wretched safe to be near, even if just for a moment. As Corbin puts it, "fecal matter was an irrefutable product of physiology that the bourgeois strove to deny."[10] But this recurrent denial also involved the projection onto the poor of that which the rich most feared, the poisonous odor of shit. For the French bourgeoisie in the early nine-teenth century, these anxieties often led to withdrawal into the domestic sphere, into family privacy, and disengagement from odoriferous public life. Later that century, improvements in privies and expansion of sewerage systems to enroll even some of the poor, along with upgraded deodorization techniques, would to an extent comfort and console the feces-averse elite.

With the rise of germ theories of disease causation late in the nineteenth century, dread of olfactory action at a distance even-tually declined, and the European sensory vocabulary dwindled. The public learned instead to shrink from any possible *contact* with microbes, and so, touch was all that mattered – perhaps not even that, just contiguity and contamination. Feces, as we have seen, remained the major inciter of consternation, since base matter seemed most likely to harbor invisible pathogens, just as it had always propagated the grossest smells.[11] The fear of feculent odor was translated into apprehension of excrement as a favored medium of microscopic disease agents, which only spurred more vigorous efforts to contain it, to send it far away through impervious pipes, where it might be sterilized and eliminated, rendered distant and invisible. The stench of shit was no longer itself morbific, but it might still be taken to signal material danger to hygienic bodies and likely contamination of sanitary spaces. In some ways, pathogenic agency had become more diffuse and less perceptible, yet it could still be mapped onto certain groups of people: the poor, the marginal, other races, foreigners, those suspected of infection, and those who refused to defecate with propriety. As Corbin concludes, infec-tious disease management now was based "on the systematic medical examination of the population as a whole" – and on meticulous inspection and registration of the contents of their excreta.[12] Regulation of shit thus might create and ratify the hygienic, safely bounded, and securely modern, if ultimately imperfect, bourgeois citizen.

"Our western understanding of civilization," according to Berlin-based journalist and historian Florian Werner, "is intertwined with the disappearance of shit; the degree of its (in)visibility signifies the position of a country on a scale of civilizational development. Functional underground canalizations and individual toilets with a lock – those are the harbingers of a modern industrial nation." Werner is convinced that "human culture is built on shit . . . because it is only when we draw a line of separation from shit, we come to know what culture is."[13] Some of us are fortunate to live in cities where our excrement can be flushed into invisibility and odorlessness in the bowels of the sewer system, making us feel decent, hygienic, and modern. "We need shit," Werner writes, "in order to do away with it and in order to affirm our sophistication. Shit is indispensable to our self-understanding as modern people." But we can never be truly rid of the fecal, despite our regular striving and straining. "The ostracized matter returns like an undead from the depths of the earth and from the unconscious; it knocks audibly against the canal and toilet that keep it imprisoned."[14] And so, the civilizing process can never end.

Laporte, Corbin, and Werner are elaborating, in a sense, on sociologist (and medical dropout) Norbert Elias's account of the social construction of shame and delicacy in Enlightenment Europe, the shaping of emotions of fear and distrust, the remolding of affect, that made up what he called the "civilizing process." Self-restraint and fear of transgression of social prohibitions had led to the "more complete 'intimization' of all bodily functions, towards their enclosure in particular enclaves," to the fashioning of the modern self.[15] Alongside the cultivation of this reticent and reserved self, a kind of introverted prudishness, the administrative state grew ever more powerful and bold, eventually claiming authority over social conduct and a monopoly on physical violence. Writing in the 1930s – that "low dishonest decade"[16] – and disturbed by atrocities in Nazi Germany, Elias saw civilization as a precarious period between barbarism and decadence. The civilizing process for a time had changed human conduct and sentiment, or "habitus," enhancing self-control and rationalizing social life, making the bourgeoisie ever more shit averse. Nonetheless, Elias observed ruefully, contemporary "civilized" societies continued to be riddled with contradiction, disproportion, and irrationality – spoiled with excrementitious stuff.

Anal Characters

As it took form, psychoanalysis fixated on the perilous development of bowel control and its dynamic role in character formation. True to its ancient roots, "analysis" thus manifested again as a symptom of defecation. At the turn of the century, while punctilious Boston bacteriologists were rushing to analyze sewage, Sigmund Freud reflected on the sensual and erotic aspects of the infant's withholding and releasing of feces, investing what might seem an instrumental act with libidinal value.[17] In bourgeois Vienna, he had observed the toddler's fascination with defecation, the pleasure taken in control and expression of excreta. It did not escape the pioneering psychoanalyst's attention that many sexual processes and organs are derived from excretory ones. But the anal zone is denied its erotogenic significance in normal development. Freud, who often was constipated, postulated that toilet training, the forbidding of sensual pleasure, and the infant's rebellion against it, might lead to various sublimations, through which illicit impulses take socially acceptable shape, and to reaction formations, defense mechanisms through which subjects manifest the opposite of their true feelings. Thus, some children developed barriers against these repressed tendencies, becoming excessively orderly and meticulous, or self-willed and obstinate, or parsimonious – reaction formations known as "Freud's triad." By orderly, he meant obsessed with cleanliness and annoyingly conscientious in carrying out small duties; obstinacy implied defiance, sometimes rage and vengefulness; and parsimony could end in avarice, the single-minded accumulation of money, an unconscious equivalent of shit. Later, in *Civilization and Its Discontents*, Freud would speculate on whether the infant's forced sacrifice of anal eroticism, its sense of guilt, might help explain the origin of civilization, a sublimation of the individual's libidinal repressions – a case of phylogeny recapitulating ontogeny.[18] The contents of civilization's discontents were excrementitious.

Freud's first contemplations of puerile anal eroticism and character development took place in early twentieth-century Vienna, which novelist Robert Musil would aptly characterize as Kakania or Excrementia. Written piecemeal from 1921, *Der Mann ohne Eigenschaften* surveys the modern city in 1913, in the last mediocre days of the dual monarchy of Austria and Hungary, *kaiserlich und königlich*, or kaka, a child's word in German for feces. Musil set his stories toward the end of Vienna's sewer boom, in a time that celebrated new flush toilets, but most of all, Kakania signified for him orderliness and blandness, conservatism and harmony. Its security and calm – surely another reaction formation – disguised underlying decay and deceit, a subterranean rottenness, a void where the soul should be. Musil had his principal character, the passively inquisitive Ulrich, a disaffected mathematician and flaneur, a man without qualities, observe: "I meant to say that just as we already have the technologies to make useful things out of corpses, sewage, scrap, and toxins, we almost have the psychological techniques too. But the world is taking its time in solving these questions."[19] Cue the entry of Sigmund Freud and his followers.

For Ernest Jones, the leading exponent of psychoanalysis in Britain, Freud's insight into infantile anal eroticism was "perhaps the most astonishing" of his master's findings, provoking "the liveliest incredulity, repugnance, and opposition."

That character traits "may become profoundly modified as the result of sexual excitations experienced by the infant in the region of the anal canal" seemed at first "almost inconceivably grotesque."[20] But Jones soon was convinced that the sexual pleasure the infant experienced in defecation might eventually be displaced into reaction formations such as orderliness, obstinacy, and parsimony – all of them barriers against repressed tendencies. These excrement-denying character traits, Jones wrote, "are closely related to narcissistic self-love and over-estimation of self-importance." Such ascetic or self-martyring impulses demonstrated a desire for continuing self-control. But the adult must repress the true object of fascination, the excretory product, substituting "faecal symbols" like children and money. The "anal-erotic complex" therefore is cathected to instincts to possess and to create replacement objects. Reaction formation to the infantile "retaining" inclination would result in a "high capacity for organizing and systematizing" and intolerance of waste.[21]

The psychological consequences of anal eroticism, the psychosexual pathologies of infant toileting, fascinated the growing band of international Freudians. New York psychoanalyst Abraham A. Brill, the first rather plodding translator of Freud's oeuvre into English, soon offered a facile interpretation of the process. In 1912, he described three wealthy New Yorkers who were obsessively neat, orderly, and stingy – all belonging "to that class of persons who prolong the act of defecation by reading books and newspapers in the water-closet." The analyst was preoccupied with a forty-four-year-old merchant who kept thinking of feces whenever he ate a sausage, dreamed of people having bowel movements, imagined toilet paper when shaking hands, and saw the moon as a rectum. "A person with protruding teeth would recall feces protruding from the anus," Brill informed his readers.[22] This anally obsessed businessman showed "extreme interest for feces and for the gluteal region," often saying "kiss my behind," which indicated "a repressed

pleasure." As a child, the poor patient, the teller of this "scatological story," had trouble with rectal control, for which he
was frequently punished.[23] As an adult, he became constipated,
offended by the slightest infraction of any rule, and exceptionally miserly. He even refused to pay Brill's costly fees.

Meanwhile in Berlin, during the Weimar Republic,
second-generation psychoanalyst Karl Abraham, supposedly
Freud's best pupil, also became concerned about the doleful
consequences of the child's "surrender of excrement." From
observation of his own offspring, he confirmed that the infant's
early coprophilia soon "proceeds along the path of reaction
formation to a special love of cleanliness." The initial "primitive feeling of power," a source of infantile "megalomania,"
once associated with evacuation of the bowels is gradually
diminished with toilet training.[24] Thus, through intense discipline, "the view of the excretions as a sign of enormous power
is foreign to the consciousness of normal adults." Instead, they
come to take pleasure "in indexing and registering everything,
in making up tabular summaries, and in dealing with statistics
of every kind." These anal characters immerse themselves "in
compiling lists and statistical summaries, in drawing up pro-
grammes and regulating work by time sheets."[25] One imagines
they also display special fondness for epidemiological analysis.

The vicissitudes of anal character continued to intrigue
social theorists through the middle parts of the twentieth century.[26] In general, critics sought to transfer the site of character
formation from the potty to social structures such as capitalism and bourgeois modernity, subduing sexual aspects of toilet
training and amplifying the influence of the civilizing process
on libidinal strivings. Thus, Freud would in effect meet Karl
Marx to join in critique of repression and alienation in modern
societies. From the 1940s, Erich S. Fromm, the German-
American psychoanalyst and sociologist, based in Mexico City
during the Cold War, argued generally that social imposition
of feelings of guilt and shame through the whole course of

development caused disunity and conflict in human identity, the play of repression and reaction formation. The child might gradually establish freedom *from* parental authority, he claimed, yet fail to acquire the freedom *to* be independent or "authentic," instead developing an authoritarian personality, which was modified anal character.[27] Fromm believed Freud's concept of anal character therefore should be regarded as a broad indictment of the norms of nineteenth-century bourgeois modernity, not just a function of family discipline.[28]

Writing in the 1950s, developmental psychologist Erik Erikson, who studied psychoanalysis in Vienna with Anna Freud, also tried to connect individual maturation and evolution with larger cultural contexts, linking libidinal character formation with socioeconomic and political structures. In *Young Man Luther* (1958), the psychoanalyst described how the constipated priest experienced an "identity crisis" while on the toilet, changing "from a highly restrained and retentive individual to an explosive person; he had found an unexpected release of self-expression."[29] Ostensibly it was the oppressive environment of the unreformed church, not the family drama of toilet training, that prompted this rebellion and resort to anal vulgarity, a character-building crisis that eventually was afforded a healthy resolution in Protestant sublimation. Around the same time, German-American social critic Herbert Marcuse proposed that bourgeois civilization, not infant development alone, was the prime driver behind the subjugation of human instincts. We moderns react to surplus repression under capitalism by becoming one-dimensional consumers and hoarders of commodities. And yet, he wrote in *Eros and Civilization* (1955), other modes of character formation are possible, less repressive styles of civilization may be within our grasp, sexual liberation is still imaginable. Anal character and authoritarian personality might yet be circumvented.[30] Ever a civilizational salient, the anus thus became one of the main intellectual battlefields of the Cold War.

American classicist Norman O. Brown showed perhaps the most intense engagement with anality and shit during the 1950s. Stimulated by conversations with Marcuse at Wesleyan University, Brown became intrigued by the "anal character of civilizations," given that "repression weighs more heavily on anality than on genitality." His bestseller, *Life Against Death* (1959), explored the "essentially anal-sadistic structure" of modern civilization, its dependence on bodily and mental repression, invoking the rise of capitalism and Protestantism as causative mechanisms to substitute for Freud's enclosed family dynamic.[31] At the same time, Brown, who went on to contribute to the history of consciousness program at the University of California, Santa Cruz, insisted on the libidinal, embodied attributes of repression and sublimation – for him, "anality means real bodily anality," not any conventional abstraction. Most contemporary social theory seemed just another sublimation, too distant from actual bodies that shit and from minds that are made to feel guilty about it.[32] Like Marcuse, Brown hoped his Freudo-Marxist analysis might offer an alternative theoretic rationale for a less repressive civilization, a resurrection of Eros, a Dionysian spirit of abandon – an aspiration soon embraced by long-haired devotees in emergent counter-cultures and serious youngsters in the New Left.[33] And so, to a remarkable extent, Cold-War liberation movements may be sourced in the rectum.

Orificial Orders and Their Contents

"Why should bodily refuse be a symbol of danger and of power?" asked Africanist social anthropologist Mary Douglas in the 1960s. Rejecting any individualistic explanations from developmental psychology, Douglas suggested that humans inherently perceive all margins as dangerous and believe all orifices are potentially vulnerable. Accordingly, the symbolism of the body's boundaries represents imagined risks to community integration and organic unity. Anxieties concerning the body's orifices have sociological counterparts in fears about security and stability, worries about exits and entrances, distress about the durability of hierarchy and order. Apprehensions of individual polluting behavior offer analogies to possible threats to social and political structures. Our embodied actions are heavy with symbolic meaning. "We cannot possibly interpret rituals concerning excreta, breast milk, saliva and the rest," Douglas wrote, "unless we are prepared to see in the body a symbol of society, and to see the powers and dangers credited to social structure reproduced in small on the human body."[34]

Brought up in a peripatetic British colonial family, Douglas had become interested in the disturbing influences and the special valences of shit, dirt, and other non-compliant and unassimilated matter out of place. "Ideas about separating, purifying, demarcating and punishing transgressions," Douglas observed, "have as their main function to impose system on an inherently untidy experience." In general, the sense of pollution or profanity, the suspicion of promiscuous defecation and wonton endangerment, tends to position imperiled innocence and purity against transgressive matter requiring removal and transgressors deserving punishment. Fears of pollution thus "brand the delinquent and . . . rouse moral fervour against him." Believing themselves at special risk, those higher in the social hierarchy turn on the alleged violators, the reckless polluters, the shameless defecators, attributing blame and conferring stigma on them. There are various bourgeois "rites of reversing, untying, burying, washing, erasing, fumigating, and so on" – all directed at separating and protecting the supposedly vulnerable from supposed contamination.[35] The work of civilization is dedicated to the goal of securitization, to distancing the pure and wholesome from matter out of place, from excrementitious stuff. Part of the prophylactic transformation involves rendering dejecta abstract and calculable, incapable of causing defilement, through "a long process of pulverizing, dissolving and rotting away," followed by further analysis and writing up, as in epidemiological surveillance. Thus, the abomination "is put into a very special kind of ritual frame that marks it off from other experience. The frame ensures that the categories which the normal avoidances sustain are not threatened or affected in any way. Within the ritual frame the abomination is then handled as a source of tremendous power."[36] For Douglas, our preoccupation with the management of shit therefore is not so much a reaction formation as a symbolic control mechanism, less about individual character formation than lending stability to social structures.

According to Bulgarian-French social theorist Julia Kristeva, Douglas was too vehement in rejecting Freudian analysis, too ready to dismiss subjective dimensions of defilement. The prim Anglo-Catholic anthropologist had focused too narrowly and too defensively on boundary maintenance, whether corporeal or sociological. This concentration on the "demarcating imperative" led to neglect of the subjective experiences of shit and other excretions as they disturb identity and disrupt system – the experience, that is, of abjection. "Excrement and its equivalents (decay, infection, disease, corpses, etc.)," psychoanalytically trained Kristeva tells us, "stand for the danger to identity that comes from without: the ego threatened by the non-ego, society threatened by its outside, life by death." "Waste" products, comprising the abject, are what we "permanently thrust aside in order to live." Excrement, urine, blood, and sperm constitute the abject, "something rejected from which one does not part, from which one does not protect oneself as from an object."[37] Much as we try to distance ourselves from shit, it always returns to discompose and vex our sense of self, as a flux that erodes enunciation. According to Kristeva: "It is thus not lack of cleanliness or health that causes abjection but what disturbs identity, system, order. What does not respect borders, positions, rules. The in-between, the ambiguous, the composite." We find joy in the abject even as we reject it; it draws us closer even as it repels us; we yearn for it even as we condemn it. For Kristeva, the abject is a "symptom," a structure lodged within the body, a "non-assimilable alien" that we constantly try to expel, lest it may destroy our identity, unnerve our character.[38] "In the symptom the abject permeates me, I become abject," she writes. "Through sublimation, I keep it under control." She describes various ways of "purifying" or transcending the abject, mostly religious and artistic – but we might add to the list statistics and epidemiology, of course, along with other defilement rites delineating symbolic orders and paternal laws and writing machines. Thus, it is abjection that "modernity has learned to repress, dodge, or fake."[39] Or to put it another way, it is abjection that makes us want to be modern at the same time it keeps denying us unadulterated modernity.

In reading Douglas and Kristeva, it sometimes seems the building blocks of twentieth-century social theory consist mostly of shit. That is, shit, like sex, has proven an inexhaustible generator of discourse.[40] Let me briefly put forward just a few more examples among the extrusions of anal philosophy. Recently, Peter Sloterdijk has reframed the excremental imagination, our obsession with matter out of place and abjection, as the longing for autonomous immune systems, indispensable when we escape from a "fecofugal way of life" and seek refuge in self-securitization and air-conditioning. The idiosyncratic German philosopher links the emergence of modernity to a desperate need for urbanites to evade their excrement. He observes: "While the nomads preserve the fecofugal dynamics of movement, farmers and especially city dwellers are fatefully condemned to a latrinocentric style of existence. For them, the spirit of the place and the law of the latrine converged." In close-knit cities, unmistakable emanations from burgeoning water closets proved "the impossibility of a secret act with no consequences."[41] Thus, "the shared air of the settled lies

under the spell of the sewer . . . The breath of the latrine dominated Old European urbanity like an iniquitous city deity."
Increasingly, however, this "merdocratic space" demanded
further privatization and neutralization, the fashioning of a
"deodorizing modernity." The modern conception of society implied sanitized "interactions among the deodorized in
the olfactory neutral space (human rights are preceded by
the zero-smell hypothesis)."[42] Civilization thus requires the
repression of shit, its distancing and deodorization, the sublimation of abjection. With neutralization of fecal emanations
and securitizing of base matter, it becomes possible to envi-
sion "a national informatic climate system whose purpose is
to ensure the large society's affective, thematic, toxic and thus
domestic self-ventilation" – a kind of sublimated mass-media
latrine. "The requirements for modernity," writes Sloterdijk,
"are fulfilled when immune relationships are . . . converted
into technical, legal and political structures," purportedly free
of shit.[43]

But as Latour admonishes us, unconsciously echoing Freud
and his not-so-merry band of psychoanalysts, we may never
have been all that modern. The anthropologically minded
French philosopher claims that the "modern" designates two
sets of practices that operate in tandem while analytically separate: one purifies or distils separate categories of culture and
nature; the other works to hybridize or confuse or taint the dis-
tinction. We moderns labor furiously to create two disparate
ontological zones, distinguishing humans from non-humans,
society from nature, law from science. We erect "a partition
between the natural world that has always been there [and] a
society with predictable and stable interests and stakes."[44] Yet
this ontological separation allows the proliferation of hybrids,
so that human society and the natural world are always getting
mixed up and jumbled through processes of translation and
mediation – which we proceed to deny or refuse. Rather than
respecting a yawning gap between human subjects and natural

objects, we keep generating quasi-subjects and quasi-objects – which we feel obliged to suppress or transcend. But if we should ever admit this, if we recognize the failure of purification rituals, if we accept the eternal return of the repressed, then our delusion of being thoroughly modern dissolves. Thus, "the moderns have to imagine themselves as different from ordinary humanity. In their hands the uprooted, acculturated, Americanized, scientifized, technologized Westerner becomes a Spock-like mutant." At the same time, "we have never stopped building our collectives with raw materials made of poor humans and humble nonhumans." We have never really become disenchanted or decontaminated moderns, whatever we like to believe. "A historical succession of quasi-objects, quasi-subjects, it is impossible to define the human by an essence."[45] Notoriously averse to considering human interiority, ever resistant to psychoanalysis, Latour nonetheless seems almost to refer here to abjection, that which disturbs identity and disrupts system, what we thrust aside to be modern yet what we cannot live without. Although preoccupied with his collectivities and networks, Latour might almost be talking about shit – the filth that becomes us, the infection within us, forestalling modernity whether we like it or not. We have never been modern; we have never been shit-free.

Even Jacques Derrida, son of French-Algerian *colons* or *pieds-noirs*, generally resisted admitting he was writing shit when positing deconstruction. For all his fascination with the dirty inexpressible voidances and secrets of the sovereign, despite his critique of logocentrism and binary reasoning, Derrida was surprisingly restrained in considering defecation and scatological rites. He urged us to recognize the challenge that indeterminable elements, remainders that cannot be absorbed or retained, recurrently present to amour propre and white mythologies – to the white man's "own logos, that is, the mythos of his idiom, for the universal form of that he must still work to call reason."[46] In his later years, increasingly concerned with "the fragility and porosity of the limit between nature and culture," he speculated on the "spectacle of a spectrality: haunting of the sovereign by the beast and the beast by the sovereign, the one inhabiting or housing the other, the one becoming the intimate host of the other."[47] Derrida became interested in digestion and incorporation, processes of assimilation and transubstantiation, but very little got excreted in his texts. And yet, if sovereignty is devouring, then deconstruction surely is defecatory, an unintended consequence of waste management, of the secret-ing of secretions. Although deconstruction is more or less instantiated in that which cannot be eaten, whatever resists digestion or absorption, the accursed supplement, in Derrida's work excrement is rarely named, even if evidently haunting several of his lectures. Perhaps what Derrida should have said is that modern white folks keep trying to expel their shit but the traces cannot be permanently erased, hanging around like a bad smell. Despite pretenses, we have never been sovereign.

The Uncanny Stool

At night in his cabin on the cruise ship *Gunnar Myrdal*, elderly Midwesterner Alfred Lambert thought he saw a crafty turd at the foot of the bed. The severe father in Jonathan Franzen's novel *The Corrections* had encountered "a sociopathic turd, a loose stool, a motormouth," a piece of excrement that tormented him. "'Me personally,'" the agential turd informed him, "I am opposed to all strictures. If you feel it, let it rip.'" But Alfred, despite his incontinence, insisted that civilization depends upon restraint, derives from controlled defecation. "'Civilization?'" the turd retorted. "'Overrated. I ask you what's it ever done for me? Flushed me down the toilet. Treated me like shit! . . . Tightasses like you been correcting every fucking word outta my mouth since I was yay big. You and all the constipated fascist schoolteachers and Nazi cops.'"[48] The talkative turd sized up Alfred as an "anal retentive type personality." Soon the turd was smearing itself everywhere. It seemed to multiply. Throughout the cabin there were "turdish rebels snuffling sneakily about, spreading themselves in trails of fetor." Horrified, Alfred felt himself "under siege by a squadron of feces." He fled to the bathroom. "There was a science of cleanliness, a science of looks, a science even of excretion as evidenced by the outsized Swiss porcelain eggcup of a toilet, a regally pedestaled thing with finely knurled levers of control."[49] But it was too late. The cheeky turd had done its worst, smearing his character, disturbing his sense of self, debasing him, shaking loose the remnants of his retentive modernity.

Few loose turds are so articulate, but most are similarly uncanny. At the end of World War I, Freud described the *unheimlich* or uncanny as something familiar that has been repressed but keeps coming back to discompose identity and disrupt sovereignty – something alienated or hidden which returns, something repugnant which comes to light again. "An uncanny experience," he wrote, "occurs either when infantile complexes which have been repressed are once more revived by some impression, or when primitive beliefs which have been surmounted seem once more to be confirmed."[50] Among examples of the uncanny are wax figures, dolls, automata, eyeballs, animated corpses – and lively excrement. They all elicit deep anxiety. All are marked by the psychic compulsion to repeat. They are the sort of thing one cannot get out of one's mind. They keep escaping repression. As we have seen, uncanny stools abounded in twentieth-century European psychoanalysis and social theory, just as they inserted themselves into epidemiological and medical sciences. Their uncanniness keeps reminding us we have never been purely modern, just as we have never been purely wastewater epidemiologists.

Implacably, the unspeakable thus excites unremitting discourse, whether sociological or medical. It is as though social theorists and medical scientists, usually white people, must perpetually act as sentinels, ever alert for even a momentary fecal transgression, a brown stain on the white background – their vigilant watchfulness oriented toward a potentially catastrophic future. Philosophers and epidemiologists ceaselessly patrol social boundaries and corporeal orifices, mapping hostile territories, shoring up defenses and immunities, protecting our borders. Excrement has acquired premonitory eloquence, a logorrhea that cannot be blocked.

3

The Colon-ized World

In Toni Morrison's novel *Tar Baby*, the main romantic lead, a Black American fugitive convict called Son, watches Valerian Street, the white American owner of a Caribbean estate, eating a meal. Son begins to reflect on the priorities of elite white people in the islands:

> They could defecate over a whole people and come there to live and defecate some more by tearing up the land and that is why they loved property so, because they had killed it soiled it defecated on it and they loved more than anything the places where they shit, would fight and kill to own the cesspools they made and although they called it architecture it was in fact elaborately built toilets, decorated toilets, toilets surrounded with and by business and enterprise in order to have something to do in between defecations since waste was the order of the day and the ordering principle of the universe.[1]

Of these latrinocentric colonizers, white Americans appeared the more hypocritical, the ones more likely to try to repress or re-channel their scatological imperialism, the more flagrant of anal characters and authoritarian personalities. Perhaps,

Son thought, it was "because they were new at the business of defecation [that they] spent their whole lives bathing bathing bathing washing away the stench of the cesspools as though pure soap had anything to do with purity."[2] It seemed to him like willful self-denial or a kind of false consciousness. White Americans in the Caribbean refused to admit what shits they are – and yet:

> That was the sole lesson of their world: how to make waste, how to make machines that make more waste, how to make wasteful products, how to talk waste, how to study waste, how to design waste, how to cure the people who were sickened by waste so they could be well enough to endure it, how to mobilize waste, legalize waste, and how to despise the culture that lived in cloth houses and shit on the ground far away from where they ate.[3]

A few years later, when novelist Jamaica Kincaid recalled Antigua, her home island, she too observed that colonial buildings, especially the educational institutions, all resembled giant latrines.[4]

While the previous chapter attempted to put the anal back into analysis, I would like here to reimagine the colon of the colonized and to reveal the delusional self-disemboweling of white colonizers. It has become clear that anxieties about the body's boundaries, and the hazards that shit and other matter out of place pose to corporeal integrity, can be expressed through cultural demarcations and social hierarchies – from body personal to body politic. We all participate in the cultural economy of waste – wasting others or getting wasted – and in so doing we acquire, or get assigned, unequal valuations within it. In this chapter, I focus on distinctions made between nicely sewered colonizers, who can discharge secretly and securely what they most fear, and those deemed unhygienic, the multiply *colon*-ized, who serve as object lesson and warning, signifying inherent danger. In making racially visible the allegedly polluting behavior of others – a delinquent excremental mode of being that is imagined as a special peril to supposedly innocent and pure white colonizers – health officers and medical scientists etched imperial power relations deeply onto bodies and behaviors. In colonial public health regimes, the subject brown and black bodies became mired in abjection, represented as excessive producers of contaminating matter, promiscuous defecators potentially transgressing the safe havens of imagined pure and shit-free white colonizers.

The colon-izing project gained force from the late nineteenth century onward, as colonial governance became more modern and intrusive and disciplinary.[5] Public health officers and epidemiologists turned their imperial possessions, their tropical territories, into an incommodious human wasteland, imagining the local inhabitants infected or brown-washed with a film of germs.[6] Once colon-ized in this way, subject races were expected to perform their abjection and then be disinfected and sanitized, taught the discipline of the latrine, in the hope that they might approach hygienic citizenship, incrementally turn anally retentive, costive – and therefore become, in effect, de-colon-ized, the realization of which would always be dubious, its ratification endlessly deferred. Accordingly, the late-colonial state was delineated on racialized bodies (white or brown or black), intimately reduced to orifices (closed or open), and dejecta (invisible and safe or visible and dangerous). Human excrement came to constitute a commanding colonial code and canon. After all, as Morrison put it, "waste was the order of the day and the ordering principle of the universe."[7]

Caribbean-born novelist V.S. Naipaul, fashioning himself as an English gentleman, expressed the animadversion, the colonial excremental vision, perhaps most blatantly. Visiting his ancestral homeland, India, he was appalled by public companionable defecation, the abundant piles of excrement, the lack of any sense of shame. "Indians defecate everywhere," Naipaul wrote. "They defecate, mostly, beside the railway tracks. But they defecate on the beaches; they defecate on the hills; they defecate on the riverbanks; they defecate on the streets; they never look for cover." There seemed to be no interest in sanitation, not even in the lavatories at New Delhi's international airport. Naipaul saw that even there: "Indians defecate everywhere, on floors, in urinals for men (as a result of yogic contortions than can only be conjectured). Fearing contamination they squat rather than sit, and every lavatory cubicle carries the marks of their misses. No one notices." Acculturated to British mores, Naipaul found the lack of modern decency disgusting and menacing.[8] Historian Dipesh Chakrabarty has noted that the novelist's hygienic perception "speaks the language of modernity, of civic consciousness and public health, of even certain ideas of beauty related to the management of public space and interests, an order of aesthetics from which the ideals of public health and hygiene cannot be separated."[9] And that – it seems to me – is just the beginning of what the novelist is saying.

Until now, I have concentrated on the salience of shit in the fashioning of modern sovereignty and in its recurrent dissolution and deconstruction, but I want to look outside the white family drama to see what attributions of uncontrolled feculence, such as Naipaul's, do to others.[10] My concern in this chapter is the projection of what often are particularly Euro-American obsessions onto different races, a transference of complex and severely imbalanced tensions between probity and precarity onto the colonized. Despite a brief Caribbean excursion, I concentrate here on the United States' colonial regime in the Philippines, the most modern of imperial apparatuses and an early model for development programs in the twentieth century.[11] In the Philippine archipelago, colonial health officers – mostly white men from the eastern seaboard of the US – used proscribed shit and prescribed waste disposal practices, appropriately datafied and inscribed, to stigmatize and discipline supposedly errant Filipino bodies, to correct supposedly filthy and harmful Filipino customs and habits. Thus, they set subject races on a trajectory that purportedly could render them toilet trained and repressed, pseudo-modern and conflicted, just like the colonizers. The ethos and infrastructure of racial development and nation formation were predicated upon the recapitulation and scaling up of child development, amplifying and broadcasting the pathos of potty training.

Excremental Colonialism

In the early twentieth century, the colonial Bureau of Health in the Philippines, run by white American health officers, did not doubt the special dangers of Filipino excrement. Human wastes, it warned in a public health bulletin, "are more dangerous than arsenic or strychnine." The excretions of local inhabitants were exceptionally heavy with minute organisms conveying dysentery, cholera, typhoid, and other diseases. Filipinos needed to treat their "evacuated intestinal contents as a poison," avoiding contact with them and abstaining from contaminating others. According to Americans in the islands, subject races seemed to lack control of their orifices. "The native and Chinese population," lamented Dr. Wallace DeWitt, "tend markedly to increase the general unhygienic surroundings by reason of their uncleanly habits." Dr. Thomas R. Marshall agreed that "the Filipino people, generally speaking, should be taught that ... promiscuous defecation is dangerous and should be discontinued." American perceptions of the local excretory regime ignored the pernicious influences of poverty, lack of infrastructure, and the effects of recent war and dispossession – rather, supposed looseness of bowel control among Filipinos was attributed to inherent cultural proclivities and retardation in civilizational progress, a racial defect. "They appear to me," reported Surigao medical officer Henry du Rest Phelan, "like so many children who need a strong hand to lead them in the path they are to follow."[12] Medicos like Phelan believed that all the promiscuous soiling was mocking and threatening to transgress supposedly closed and fixed colonial boundaries. Even worse, this reckless pollution behavior in the tropics appeared to endanger pure and innocent white Americans far more than the defiant brown transgressors. Filipino shit harbored all sorts of pathogens to which the white foreigners, previously unexposed, felt uniquely vulnerable – even when sickness and mortality rates suggested otherwise.

A grotesque defecating Filipino body regularly irrupted into colonial medical reports and epidemiological studies. For example, Philip E. Garrison claimed he had discovered "one of the most striking instances in the history of medicine of a population almost universally infested with animal parasites." He therefore recognized the "imperative need" to regulate the "methods of the disposal of excreta customary among the Filipino people."[13] In a cholera outbreak, military surgeon Edward L. Munson found that asymptomatic Filipino vibrio carriers "would seem not only the most numerous but the most insidious and dangerous sources of infection." Unable to source enough samples from inadequate Manila sewers, Munson sent out squads of sanitary inspectors:

> The work meant invasion of the accepted rights of the home and the individual on a scale perhaps unprecedented for any community. The collection of the fecal specimens necessary might fairly be regarded as repulsive to modesty. Add to this the facts that the search was made among persons apparently healthy to themselves and others who could scarcely fall even within the class of suspects. . .[14]

In 1909, the hard-pressed white laboratory workers at the Manila Bureau of Science had examined more than 7,000 fecal specimens, almost all from Filipinos; and then in 1914, at the start of the cholera epidemic, they were deluged with more than 126,000 jars of Filipino excrement.[15] The civilizing mission would be long and dirty, it seemed. An American physician declared: "the cleaning of the Augean stables was a slight undertaking in comparison with purifying the Philippines . . . No imagination can make the Filipino customs with respect to [defecation] worse than actuality."[16] According to medical authorities, such uncivilized colon-ized subjects would need generations of instruction in the proper technique of evacuating their bowels, inculcation in the habit of secreting excrement, before inherent tendencies to communicate freely their feces might be repressed or eradicated.

For American medical officers, the Filipino homunculus was dominated by the anus; whereas the expressive and devouring mouth generally symbolized white presence. Reading medical reports and scientific treatises, one gets the impression that Filipinos did nothing but shit uncontrollably, making waste; while white American men remained retentive, or at least managed to discharge secretly and safely, even meticulously, their innocuous excreta. The colonizers spoke, wrote, surveyed, experimented, policed, supervised servants, hunted, fished, and fought; they engaged in sporting contests; they collected stuff obsessively; but they appear not to have gone to the toilet. They relentlessly expressed an American sublime – that "sublimated spectacle of national empowerment"[17] – even as their repressions, or rather reaction formations, lent force to colonial violence and domination. It was yet another operationalization of the warlike "ascetic body," extoled in rising US imperialism.[18] White Americans, so modern, so civilized, thus fantasized that they could be alienated, like automata, from the filthy exuberance and messy relationality of the tropics, avoiding close encounters of the turd kind. But even as they denied it, Americans still were fascinated by defilement and the boundaries it marked in such a vulnerable manner; still haunted by the abject, which could never be permanently expunged; still forced to submit to having repressions return to bite them in their disowned fundaments. No wonder so many of these "masterful" white men broke down, succumbed to brain fag and neurasthenia, went troppo.[19]

An Archipelagic Laboratory?

If the colon-ized were not ready for modern sewer systems, they could at least accommodate themselves to pits and latrines. "Let those who are able to put in septic tanks and flush closets do so," declared the Bureau of Health – all others must arrange a "sanitary pail" system by digging a pit and covering their excrement with lime or clean soil.[20] Public health experts recommended raising a wooden frame with a self-closing seat above the pit, which contained a large can. This meant the "container for the can has the advantage of being entirely open, which fact secures good ventilation and leaves no opportunity for the collection and retention of disagreeable odors." For a fee, a carabao cart would regularly haul away the cans, delivering them securely to "a suitably centrally located pit." Even so, many years passed before the pail system was widely accepted. Poorer barrios in Manila depended on a few clusters of public "insanitary closets," or none at all.[21] It was much worse in the provinces. When Dr. David Willets arrived in the Batanes Islands, he observed that "a suitable method for the disposal of human excrement is lacking." Latrines or water closets of any kind were exceedingly rare – "and furthermore the people have not learned to use them."[22] He therefore organized disinfecting squads that eagerly sprayed carbolic acid over people and their dwellings, as well as liming everything that displayed any trace of fecal matter. Having failed to toilet-train recalcitrant Filipinos, especially those lacking toilets, Americans would frequently resort to spraying carbolic and spreading lime.

By 1920, urban Filipinos were enjoined to obey the new sanitary code, which required all buildings to include "adequate privies or toilet accommodations, constructed according to plans approved by the Director of Health." Sanitary inspectors demanded to see, as the bare minimum, "a pit not less than one and a half meters in depth, securely covered by a slab of stone or concrete . . . a seat, provided with a cover, so devised to close automatically when not in use; a vertical conducting pipe . . . leading from the seat to within the pit; and a vent pipe not less than ten centimeters in diameter leading from the pit to one meter above the eaves of the building." This "Antipolo toilet" – named for the Catholic pilgrimage site outside Manila – boasted a capacity of one cubic meter for each Filipino resident, more than deemed necessary for white Americans, who presumably were less extravagant defecators. The sanitary code emphasized the importance of not allowing the Antipolo toilet to "communicate" with any other room, and to possess a "tight-fitting door."[23]

Colonial tropes of purity and danger, of refinement and defilement, were not only racialized – they were spatialized too. Health officers thus campaigned to render unsanitary markets as toilet-like as possible (and hence sanitary, somewhat paradoxically). In the white American imaginary, traditional markets in the Philippines were exceptionally filthy and disturbing, places of promiscuous contact and uncontrolled contamination, areas of darkness where colonial hierarchies may be suspended. They seemed naturally perverted spaces. James A. LeRoy, a colonial bureaucrat, was sure that "unless there be rigid and efficient supervision," the markets inevitably were "foci of infection." For lawyer Daniel R. Williams, the markets were nothing but "unwholesome and death-dealing plazas." "No one who has not traveled in the Orient can conceive of the noise and confusion," teacher William B. Freer wrote of Manila's street-life. "Words fail utterly to describe it."[24] The only solution was to construct new markets of hygienic reinforced concrete and to regulate commerce within the sterile edifices. These modern concrete markets, erected throughout the archipelago, became, in the words of the authoritarian and frequently constipated director of health, Victor G. Heiser, "educational features . . . doing much to spread the doctrine of cleanliness throughout the islands."[25] When he departed Manila, on his way back to New York to become the director for the East of the Rockefeller Foundation's International Health Board, Heiser wrote: it was "a satisfaction to see the indestructible monuments of cement which I left on the landscape and which they will be unable to destroy."[26] The author of *Interesting Manila*, travel writer George Amos Miller, recalled that "before the days of American sanitation, the condition of [the markets] was indescribably bad, but modern regulations and efficient inspectors have changed all this to comparative cleanliness and good order." Journalist Frank G. Carpenter remembered that in 1900 the largest marketplace in Tondo "consisted of ten acres of rude sheds, roofed with straw matting or galvanized iron laid upon a framework of bamboo poles."[27] But by 1920, it was a well-ventilated building of concrete and steel, hosed down and flushed out every night – just like a giant cemented latrine.

Places of defecation and locations of commerce – toilets and markets – were meant to resemble modern scientific laboratories, where Filipinos ideally might perform everyday life as respectably as white American laboratory workers. The medical laboratory had become the exemplary locus of colonial modernity, both sign and signifier of difference.[28] Tropical laboratories established in Manila after 1901 developed a distinctly "abstract" and de-libidinized spatial texture, an alienated and ascetic topography, characterized by isolation, disinfection, hygiene, standardization, and regulation.[29] They were constituted as places of bodily cleansing, sterilization, white coats, correct training, rigid hierarchy, and avoidance of contamination. As the laboratory inscribed and abstracted the world around it, so too did its spatial practices ensure the young white men making their discursive mark within it were able to transcend their own lower bodily stratum and set themselves aside from the filth around them, at least for a moment. It was like what Friedrich Nietzsche had observed a few years earlier, when he deplored the calculability of modern science and the concomitant tribulations of distance: "It is necessary that the emotions be cooled, the tempo slowed down, that dialectic be put in place of instinct, that seriousness set its stamp and gesture – seriousness, which always bespeaks a system working under great physiological strain."[30] Obedience to these demands might enable the laboratory and its homologs, the latrine and the sanitary market, to serve as models or aspirations for modern governance. And surely, they did all involve strain – straining against stool, metaphorically at least. "The Philippines may be considered today as a laboratory," LeRoy announced hopefully.[31] He endorsed efforts to render everything in the unruly dirty tropics more laboratory-like, whatever the cost. According to Bruno Latour, pioneering germ theorist Louis Pasteur in the late nineteenth century also had sought to "enroll" humans and non-humans, including microbes, in his science, to make the laboratory central to their way of being in the world. Society, Latour tells us, is assembled as the result of passage through such laboratories. "In this succession of displacements, no one can say *where the laboratory is* and *where the society is*."[32] Thus, no one can say where the laboratory ends, and the modernizing colonial state begins. Or where the latrine ends, and modern citizenship commences.

Late in the 1980s, as I wandered through the remains of the colonial Bureau of Science and the Philippine General Hospital, those peculiar mishmashes of dirtied sterility and Spanish ornamentation, of polished wood and borer, of degraded concrete and corroded metal, my mind kept turning to the notion of tropical kitsch. Not the kitsch of consumerist art commodities, but kitsch in the sense of that which is mechanical and formulaic, imitative and uncritical, associated with certainty and closure, not irresolution and profanation.[33] There was something rehashed and hackneyed about the ruins of the colonial laboratories, something that resisted local embeddedness, refused vernacular relations, something distracting and diversionary about these failed projects. They were heavy with a decorativeness that avoided, or at least displaced, all the unpleasantness and misery that once had seemed to justify them. One hundred years later, they still were striving in vain for a simple stereotype, a suitably reductionist ontology. The decaying edifices reminded me of what Saul Friedlander called "uplifting kitsch," a kind of stylization of the values of a group that might serve to mobilize the population, kitsch as a mechanism of cultural control. In studying Nazi kitsch, Friedlander found that forms relating to death scenes, debility, and sacrifice exerted an emotional impact that the celebration of life lacked.[34] When I imagined these colonial laboratories as tropical kitsch, I was thinking through similar attempts to estheticize politics, although perhaps less successful ones.[35] As I wandered among the colonial scientific remains in the late 1980s, I was thinking too of a book I had read not long before, *The Unbearable Lightness of Being*. Milan Kundera writes: "Kitsch is the absolute denial of shit, in both the literal and figurative senses of the word; kitsch excludes everything from its purview which is essentially unacceptable in human existence."[36] Just like the colonial laboratory, I thought. And the same may be said, too, of colonial latrines.

At this point I expect some readers will be asking what this story has to do with anyone elsewhere – say, in North America? Well, with the increasing "Filipinization" of the colonial bureaucracy in the second decade of the twentieth century, returning white health officers began their long march through institutions in the continental US, carrying with them the lessons learned through managing what they imagined to be the unacceptable excretory practices of subject races. There are many examples of the empire coming "home"; let me give just a few. William E. Musgrave, the director of the Philippine General Hospital, became professor of tropical medicine at the University of California's medical school, proffering advice to Bay Area public health authorities.[37] Louis Shapiro applied lessons learned on Bontoc, especially those regarding racial hygiene, as leader of the Milwaukee public health department.[38] As deputy director of public health in the Philippines, Allan J. McLaughlin had warned against the dangers of promiscuous defecation by inferior races; as commissioner of health in Massachusetts, he completely reformed disease surveillance and expanded efforts to enforce principles of personal hygiene, along colonial lines. Immigrants and non-whites were particular targets. No surprise, then, that McLaughlin often expressed his commitment to fecal sampling and sewage bacteriology in Boston, especially in poorer and more diverse parts of the city.[39] The imperial origins of epidemiology and public health should not be forgotten. We may never have been modern, but most of us have been colon-ized, whatever our location.

The Latrinoscene

The twentieth century became the century of the freestanding latrine. When Heiser visited the Philippines again in 1925, he was delighted that Filipinos were "voluntarily building large numbers of latrines."[40] Admittedly, they did not always turn out as a white American like himself might hope. "The question of superstructure is left entirely to the householder's wish and it is amazing to see the numbers of directions into which this feature develops."[41] More disturbing, Heiser believed that brown races still were not defecating with propriety. He instructed the local Rockefeller Foundation emissary, Dr. Charles H. Yaeger, to modify the toilet design "to make it impossible to sit on except in the desired position."[42] Across Luzon, dutifully boring thousands of latrines, Yaeger was trying desperately to drum up enthusiasm for proper defecation. "The spectacle of boring and particularly of blasting is one which appeals strongly to the people . . . This appeal to the imagination is an important aid in attempting to persuade a community to install a large number of latrines. There is nothing dramatic about the old pit latrine."[43] But he and his medical colleagues often met with annoyance that more pressing needs had been ignored. "In one instance, too much insistence on latrine installation resulted in an anonymous letter threatening the life of the district health officer."[44] It was a hazardous job, but someone had to do it. At other times, Yaeger wondered if he was subjected to parody. In one town, a local artist insisted on making a "wood carving of someone boring a latrine and suggested me. Well, a joke is a joke, and I didn't know if they were serious or not but took it in good spirit. What a reputation!"[45]

From the beginning, the Rockefeller Foundation was dedicated to making the developing world safe for the latrine. Its chief concern became the prevalence of hookworm, the "germ of laziness," in the southern US and in the tropics.[46] In the 1890s, Charles Wardell Stiles identified hookworm in the stools of poor whites in southern states, describing how the nematode parasite entered through the skin, eventually lodging in the intestines, where it fed from the bloodstream. Blood loss produced symptoms in the host of pallor, weakness, and fatigue. At the turn of the century, Bailey K. Ashford found that a new type of hookworm, later called *Necator americanus*, abounded in the feces of inhabitants of Puerto Rico, recently occupied by the US. Ashford led a vigorous campaign against the intestinal parasite on the island, setting up mobile field hospitals, administering a nauseating thymol mixture, and emphasizing safe stool disposal. Impressed with this colonial model, the Rockefeller Sanitary Commission established similar programs in each southern state, focusing on diagnostic investigations of fecal specimens, mass dispensing of thymol, and educational crusades against "soil pollution." Rockefeller agents traveled from town to town, putting up scary displays of hookworms and other helminths, showing model sanitary

houses and latrines, exhorting the public to fend off the excremental germ of laziness – it was a style of hygiene evangelism
that echoed old-time southern tent revival meetings. Black
defecation always seemed especially menacing to vulnerable
poor whites. As Stiles put it, "the white man owes it to his own
race that he lend a helping hand to improve the sanitary surroundings of the Negro."[47] Charles T. Nesbitt, a public health
leader in North Carolina, was blunter: "The hookworms, so
common in Africa, which are carried in the American Negroes'
intestines with relatively slight discomfort, were almost
entirely responsible for the plight of the southern white. It is
impossible to estimate the damage that has been done to the
white peoples of the South by the diseases brought by this alien
race."[48] Nesbitt was convinced that until African Americans
could be induced to defecate correctly, their intestinal contents must be rigorously segregated, treated as poison, kept at
a distance from valuable whites.

Impressed with sanitary progress in the southern United
States, the Rockefeller Foundation's International Health
Board scaled up the hookworm campaign to encompass the
colonial tropics. Hookworm programs offered the Foundation
an appealing means of ingress into otherwise closed imperial
possessions and new dominions, first in Egypt and the tropical north of Australia (where Aboriginal, Asian, and Pacific
Islander feces were execrated), then throughout Southeast Asia
and the Pacific, with Heiser taking charge.[49] During the early
twentieth century, thousands, then millions, of latrines came to
dot colonial landscapes and ring the equator. Across the globe,
Rockefeller men like Yaeger relentlessly bored holes for ever
more latrines. Most of the installed privies, however, proved
far less sophisticated and much cheaper than those favored in
the continental US.[50] After World War I, the latrine craze soon
infected the new League of Nations' Health Organization, only
it would be called "social medicine" or "rural hygiene" in the
interwar years.[51] Environmental sanitation became part of a

suite of measures that ostensibly addressed the socioeconomic causes of disease, the underlying structural etiologies of the developing world's burden of illness, without disturbing the global capitalist system. Latrine construction was indelibly associated with modernization, just as confined and precise defecation was with hygienic citizenship – or at least with eligibility for such citizenship. After World War II, the World Health Organization continued the theme of regulating human waste disposal, especially in Asia, though it concentrated on the laying down and extension of urban sewage systems – even so, latrines were not forgotten.[52] Through the World Health Assembly, Abel Wolman, the professor of sanitary engineering at the Johns Hopkins University, emerged as a dedicated advocate of sewering the great unwashed of the world, just as he had sewered the poorer parts of Baltimore.[53] International organizations in the twentieth century thus ensured that the latrine and the sewer became key technologies working both to protect and secure white people and to extend the boundaries of whiteness, to render darker races almost "white," but not quite.[54]

One member of the Rockefeller family, however, came to regard the sewer, particularly its sludgy residue, as part of the problem – imperiling the virtues of the autonomous latrine and therefore just as threatening as any promiscuous defecation. In the early 1990s, while I happened to be writing about excremental colonialism, a neighbor in Cambridge, Massachusetts, invited us to a party where she would unveil her new composting toilet, ensconced in the house where the poet E.E. Cummings had been born – and a few doors from where William James, almost 100 years earlier, had written on matter out of place. I could have been walking into a scene from Don DeLillo's *Underworld*. An avowed Marxist and radical feminist, Abby A. Rockefeller imagined she was diverging from the family legacy, though I could not resist the thought that her relentless promotion of latrines – albeit incorporating on-site aerobic composting systems or anaerobic digesters – remained compatible with enthusiasms of previous generations. Rockefeller did condemn her family's favored pit latrine as not "environmentally viable"; but she saved her fiercest reproaches for the sewerage systems producing noxious sludge. Excremental sludge, she believed, "contains constituents that are hazardous to life," despite all the mechanisms of purification, all the filtration and treatment, it goes through. "Disease – from viability and regrowth of human pathogens in raw sludge, and other diseases caused by these sludge composting processes – is of major concern." Moreover, "sludge provides perfectly the conditions for combinations of thousands of chemicals to cause a cataclysmic devastation of life." Therefore, Rockefeller had decided to commit her fortune to building self-composting latrines around the world, to "protect life chains from the potential of devastation by the constituents of sludge."[55] As neighbors, most of us personally deft in the avoidance of shit, we were slightly bemused by the implacable return of excrement and the irresistible lure of the latrine to expunge it. I recall we were inclined to dismiss it as a slightly embarrassing, if utterly understandable, family trait. The composting latrine may be an ecological solution to the disposal of human waste, yet we believed its celebration that day was a symptom of deep-seated fear of something else, unnamable.

Abstracted Fecal Labor

It would be a mistake to conclude that there is something inherently wrong with care of the body, with deliberate and conscientious sanitation, and with the development of excretory infrastructures. Obviously, enteric pathogens have always contributed to the global burden of disease, often causing early deaths, especially among poor and marginalized populations. But it is far from certain that colonial public health initiatives, and their associated developmental regimes, in the Philippines and elsewhere in this period, did much to reduce rates of diarrhea and dysentery among those colon-ized – nor among the colonizers. In any case, arguments derived from such medical materialism surely could not alone explain or justify the obsessions of white administrators, and later elite nationalists, with excrement and all it symbolizes. Epidemiology alone cannot account for their guiding passion for dangerous, yet captivating, shit. For them, human wastes could be fashioned into powerful tools, stigmatizing and racializing others. Thus, the excremental vision of colonial public health, conjuring a tight race–defecation nexus, revealed new paths to racial dominance and the cultivation of whiteness.

As we have seen, late-colonial public health, in liberal humanitarian mode, worked to produce and reproduce in parallel both formalized bodies and modernized abstract spaces, positioning them against the promiscuously defecating colon-ized, represented as crapping all over the map. Elite white corporeality was erased or sublimated, while the colon-ized became the principal sources and transmitters of contaminating base matter, transgressing colonial boundaries, jeopardizing innocent and pure colonizers. According to Aimé Césaire, African poet and politician, from the colonizers' point of view "the human heartbeat stops at the gates of the Black world . . . we are walking manure."[56] Colon-izing a population in this way, wasting the population, took endless industry and striving, constant surveillance and laboratory testing, unyielding training and disciplining of subject peoples, and relentless redesign and reterritorializing – that is, development – of occupied environments. By enforcing a sealed orificial order, public health officers might bring about a seamless reformation of supposedly grotesque, typically polluting, colon-ized bodies, as well as cleansing and sterilizing the public and private parts of the colon-y. This civilizing mission, this preaching of the gospel of personal hygiene, also required colonizers to labor inexorably and intolerably toward their own repression and self-control, a task they bore with great seriousness and strain, often unsustainably so.

In this process, the stool became the power-object of the social system, something that appeared to possess commanding efficacy in determining value. I am suggesting here that feces were appropriated and fetishistically materialized, alienated from shitting subjects, fashioned as objects of sensuous desire, thereby reifying exploitative social relations.[57] As historian William Pietz, a student of Norman O. Brown, put it: such objects "are fetishes insofar as they have become necessary functional parts that are privileged command-control points of a working system of social production."[58] The attribution of inherent power to excremental stuff, its fetishization, might disguise or transubstantiate the political economies of colonial dominance and global capitalism, thus objectifying any surplus value squeezed out in the latrine. In this economy, others' feces function as an ambiguous referent, both repellent and attractive, both dangerous and remedial, available for capitalization.[59] Hence the fascination of white colonizers with the shit of others, their deeply felt need to add value to it, or at least revalorize it, in the laboratory and to regulate its circulation and exchange, to unveil it and to render it invisible, to see it and not to see it. Through such abstracted fecal labor, and the consequent alienation and commodification of excrement, colonial social relations acquired an aura of ghostly objectivity, of haunted rationalism, of humanitarian benefit – a sense of normal calculability, of ontological reduction, embracing all aspects of life.[60]

It may be, as some social theorists have speculated, that the civilizing process is epitomized in such histories of fetishization and misrecognition.[61] If we see defecation by the colon-ized as a forced labor process, with the colon-ized as workers, then the latrine becomes a factory and the stool its fetishized product. Through this defecatory labor process, the colon-ized might be revalorized, dis-credited and indebted, while the civilizing colonizers gain credit, even profit, secretly accumulating excremental surplus value. Or so they liked to believe.

4

Powers of Ordure

At the end of the nineteenth century, novelist and Spanish-trained physician José Rizal offered a rhapsodic affirmation of a fiesta in the Philippines. In the ironically titled *Noli me Tangere* – touch me not – Rizal described the turmoil and ferment in the *barrio* on the eve of the festival, the jostling and hustling, that rough and tumble, when "the air is filled with the explosion of fireworks and the blare of music. The whole atmosphere is pervaded and saturated with rejoicing."[1] The "first Filipino," the iconic nationalist, observed: "Chinese, Spaniards, Filipinos – all dressed in European or native attire – were bustling about the streets. They walked in confusion, elbowing and pushing one another: the servants carrying meat and poultry, students clad in white, men and women risking being run over by carriages and calesas." It was a time of conviviality and promiscuous contact, not at all threatening. On the day of the carnival, the whole community gathered to watch the parade, singing, dancing, eating, and drinking. A staunch advocate of personal hygiene, Rizal nonetheless was able to rejoice in "the orgy, the bacchanal to drown the lamentations of everyone!" The festivities smelled of "burnt powder, flowers, incense, and perfume; bombs, rockets and firecrackers made the women run and scream, the youngsters laugh . . . The mixture of light and motley colors attracted the eyes, and the din of harmonies thundered in the ears." During the procession, "they pushed each other, pressed upon each other, trod on each other." As Rizal recollected, "one could hardly breathe; the air was warm and reeked of human animal stench" – and yet, he did not really object to the heavy odor of sweat and slight whiff of ordure, not in these circumstances.[2]

But Mrs. Campbell Dauncey, a famously fastidious white woman, gained a completely different impression of the fiesta in 1904 commemorating the execution, by antecedent Spanish colonial authorities, of Rizal. The crowds "swarmed out" onto the streets of Iloilo in the evening. "They hang out flags and lanterns," she reported, "and every Filipino knocks off what little work he ever does, and crawls about on the streets and spits . . . while the women slouch along in gangs with myriads of children." These "ghastly" carnivals reminded her of the propensity of Filipino children and young adults to play "some mysterious, meaningless game, revolving around certain heaps of manure and dead dogs."[3] She felt besieged in the Philippines by such wanton infantile contact, her colonial haven teetering on the brink of transgression and contamination.

Modern efforts to disseminate a civilizing mission, to spread the gospel of hygiene, to diffuse the disciplines attendant on colon-ization, to develop the world, might be diverted or sabotaged at any moment by the materialization of the carnivalesque and grotesque – which Rizal celebrated and Dauncey dreaded. Filthy rites and chance wretchedness constantly threatened to undo attempts to thrust aside excrementitious stuff, subverting struggles to render innocuous the base matter that troubles boundaries and disturbs order, nullifying endeavors to erase the lower bodily stratum. I want to trace here this return of the repressed, this recrudescence of the cloaca, this insistent romancing of the turd, wending, as I do so, a path from early modern observances of the carnivalesque and grotesque, through imagined scatalogics of postcolonial governance and anthropological recovery of "excremental citizenship," to the propagation of "shit art."[4] In this chapter, I seek the recursive anus, the rebarbative stercus, the abject would-be escapee. That is, I want to look more closely at those antithetical components of the "dialectic of the sphincter":[5] not retention, but release; not purity, but danger; not immunity, but contamination and infection – whatever cannot be contained and eliminated through privy or latrine, whatever cannot be sublimated and written off. The shit that keeps coming back to haunt us – or better, that persistent stain on our undergarments, the one we cannot, or simply do not, wash away.

But this is not the usual story of resistance to constraint, the art of discarding inhibitions; it is not a liberatory narrative. There is no easy refusal of our classificatory grids, no overthrowing of system, no evacuation of abjection, certainly no ecstatic de-colon-ization of the self or de-fetishizing of the stool. Instead, I concentrate on how matter out of place is mobilized, not just repressed or sublimated, within modernity; how, when so vitalized and re-valued, it actually can give perverse force and intensity to antagonistic processes of deflection and evasion; how its expression can cruelly ennoble at the same time as it degrades. Grotesque bodies and uncanny turds roam the margins, liminal and virulent – they disrupt, yet also reconfirm, our ways of organizing ourselves and the world.[6] They both repudiate and invoke our classificatory systems; they menace conventional distinctions and standards while infusing them, below the belt, with a kind of perverse energy. As we shall see, the grotesque, like the abject, thus represents a paradox, inasmuch as it recurrently dissolves and then excitingly reconstitutes differences between the strange and the recognizable, between heterodoxy and orthodoxy.

Excremental Literary Visions

Early modern appreciation of scatological revelry, a fascination with monstrous bodies and their defiant excreta, was expressed in literary form as "grotesque realism."[7] An emerging European obsession with the lower bodily stratum was fixed on "acts of defecation and copulation, conception, pregnancy and birth" – principally, that is, on excretion. According to Russian literary theorist Mikhail M. Bakhtin – who spent the 1930s dodging Stalinist labor camps – the anus came to be presented in emerging modernity as a "deathly obstacle to ideal aspirations. In the private sphere of isolated individuals, the images of the bodily lower stratum preserve the element of negation while losing almost entirely their positive regenerating force." Thus, the developing bourgeois private sphere became liable to besmirching and degradation. The ideal modern body, coming into being, was supposed to be finished and complete, strictly bounded and enclosed, making up an individual whose every orifice was closed, imperforate.[8] Yet rebels like the writer François Rabelais posed instead monstrous and excessive bodies, transgressing any limits, anally incontinent, nothing except apertures. There were no fixed boundaries between such grotesque bodies and the world, everything interpenetrated and interacted. "The stress," Bakhtin wrote, "is laid on those parts of the body that are open to the outside world, that is, the parts through which the body itself goes out to meet the world." Such incomplete bodies, juxtaposed against early modern closed bodies, appeared to enjoy especially the sensual collectivity on display in carnivals, the clowning and foolishness, the suspension of hierarchy. "In the atmosphere of Mardi Gras, reveling, dancing, music were all closely combined with slaughter, dismemberment, bowels, excrement, and other images of the material lower bodily stratum."[9] For misfits like Rabelais, it came down to celebrating the gaping anus, a figuration opposed to emerging modernity.

Writing in the provinces of sixteenth-century France, Rabelais had discerned rich potential in grotesque debasement and the carnivalesque. A failed priest, retrained as a physician in Lyon, he wrote the four (possibly five) books of *Gargantua and Pantagruel* (1532–52) as he moved unhappily around recently consolidated France. The two gigantically gross figures in the narrative delighted in farting and shitting and desecrating on their own journeys. As an infant, Gargantua "pissed in his shoes, shit in his shirt, and wiped his nose on his sleeve: He did let his snot and snivel fall in his pottage, and dabled, padled and slabbered every where." After "dunging" behind a bush, the giant expatiated on wiping his behind, in celebratory "wipe-bummatory discourse."[10] He concluded that "of all torcheculs, arsewisps, bumfodders, tail-napkins, bung-hole cleansers, and wipe-breeches, there is none in the world comparable to the neck of a goose." Later, observing the clown Panurge, a knave and libertine, shit himself in fear, Pantagruel demanded to know: "Do you call this what the *Cat* left in the Malt, Filth, Dirt, Dung, Dejection, faecal Matter, Excrement, Stercoration, Sir-reverence, Ordure, Second-hand-meat, Fewmets, Stronts, Scybal, or Syparathe?" – and then he ignored the mess and proceeded to drink heartily. Elsewhere, Rabelais in an aside noted that monks are shunned because they "eat the ordure and excrements of the world, that is to say, the sins of the peoples, and, like dung-chewers and excrementitious eaters, they are cast into privies and successive places, that is, the Convents and Abbeys, separated from Political conversation, as the jakes and retreats of a house are."[11] As Bakhtin surmised, Rabelais and his kind "conceived of excrement as both joyous and sobering matter, at the same time debasing and tender."[12]

In contrast to Rabelais, Jonathan Swift's scatological musings were grim and scathing, though they, too, were predicated on shit's diagnostic acuity. An Anglo-Irish satirist and Anglican clergyman, Swift's "excremental vision" was expressed through numerous poems and in his novel *Gulliver's Travels* (1726).[13]

Following ship's surgeon, and later captain, Lemuel Gulliver on his picaresque adventures, Swift vividly revealed the inherently "anal character of civilizations," ridiculing vain attempts at sublimation.[14] On his early voyages, Gulliver repeatedly protested his own cleanliness: tied up in Lilliput, he assured readers, "my constant practice was, as soon as I rose, to perform that business in open air, at the full extent of my chain." But such claims of innocence could not disguise his preoccupation with base matter; he protested too much. On his third voyage, Gulliver learned from a professor at the Academy of Lagado, an expert on uncovering plots, that he must come to know the suspects' diets, where they eat, on which side they sleep, and "with which hand they wipe their posteriors." Moreover, he should "take a strict view of their excrements, and, from the colour, the odour, the taste, the consistence, the crudeness or maturity of digestion, form a judgement of their thoughts and designs; because men are never so serious, thoughtful, and intent, as when they are at stool."[15] On a subsequent expedition, abandoned on the island of the Houyhnhnms, a little south of what is now called Australia, the erstwhile hygienic captain encountered the dirty, bestial Yahoos, who threw their "excrements" at him. Producers of copious waste, the Yahoos appeared as brutes in human form. And yet, for all their "strange disposition to nastiness and dirt," these degenerates resembled Gulliver and his fellow countrymen. Disconcerted, Gulliver realized "I was an exact Yahoo in every part, only of whiter colour, less hairy, and with shorter claws." He could not get the thought out of his head. "When I thought of my family, my friends, my countrymen, or the human race in general, I considered them as they really were, Yahoos in shape and disposition, perhaps a little more civilized."[16] Gulliver found refuge with the noble race of horses, who regarded him with disdain as a somewhat elevated Yahoo. These Houyhnhnms spoke little and seemed never to shit, which is surprising for horses. On returning home, Gulliver sought out the company of implicitly Irish horses, avoiding the "English Yahoos," those degraded imperialists, whom he found repulsive.

As a critique of sovereignty, however, Swift's excremental vision, particularly his denigration of the Yahoos, is wanting in cogency. In demonstrating the degradation and grotesquery of humans, it reiterates conventional binaries, albeit with disruptive intent. According to literary critic Stephen Greenblatt, it comes close to "*nostalgie de merde*," where satiric humor still assumes, or even requires, the condition of oppression or repression, merely illuminating, or even reinstating, the impotence of the conquered.[17] After all, the dominant remain dominant, no matter how discomforted. Behind all the repugnance for shit lies durable tolerance of disorder and contamination, even fascination with that which must be forbidden, thus serving in the end to prove the immunity of the elite. Enchantment with the grotesque and the carnivalesque, although they are despised and disavowed as acceptable social forms, is "instrumentally constitutive of the shared imaginary repertoires of the dominant culture."[18] Embracing the lower bodily stratum is not so much transgressive as a licensed mode of release. Rather than fomenting revolution, carnival revelry might dwindle into "temporary retextualizing of the social formation that exposes its 'fictive' foundations," without doing much else, except perpetuate a convenient social dialectic.[19] As we know, shit has no odor when written; literary tropes do not provoke visceral disgust – if anything, they ward it off.[20]

Nonetheless, many European authors have held to the belief that defecatory enactment might intimate a critique of modernity and imperial sovereignty. Consider for a moment James Joyce's *Ulysses* (1922) and its treatment of "cloacal imperialism" – written as the Rockefeller Foundation was rolling out its latrines across the world.[21] In the Aeolus chapter, an Irish blowhard, Professor Hugh MacHugh, discourses long and loud on the "grandeur" that was Rome: "The Roman, like the Englishman who follows in his footsteps, brought to every new shore on which he set his foot . . . only his cloacal obsession. He gazed about him in his toga and said: '*It is meet to be here. Let us construct a watercloset.*'" After a few more beers MacHugh concludes defiantly: "The closetmaker and the cloacamaker will never be lords of our spirit."[22] A Rabelaisian reversal, perhaps, of Swift's excremental attributions, which once had dirtied English colonizers, but to what effect? Often read as Joyce's endorsement of Irish coprophilia against English asceticism and denial, and therefore as an anticolonial text, *Ulysses* (and similar scatological musings) also may be interpreted as restating colonial discriminations. Not so much the triumphant return of the repressed, then, as the symbolic repetition of oppression.

On Not De-colon-izing

We should bear in mind these excremental European literary traditions when reflecting upon Achille Mbembe's provocative account of the "banality of power and the aesthetics of vulgarity" that supposedly drive governance in sub-Saharan Africa after formal independence. In Cameroon, for example, "the grotesque and the obscene are two essential characteristics that identify postcolonial [post-independence] regimes of domination."[23] Drawing on Bakhtin's analysis of the Rabelaisian grotesque and carnivalesque, Mbembe argues that African regimes routinely improvise excess and disproportion, displaying crudeness and indiscipline, to assert power. For Bakhtin, the excremental and obscene resisted the arts of domination; whereas for Mbembe, they constitute modern African modes of rehearsing hegemony. The African political theorist claims that frequent recourse to the lower bodily stratum informs a "logic of familiarity and domesticity," which stimulates a "convivial tension" between rulers and their subjects – leading to "zombification," the diminution in vitality of all concerned.

Excessive performances of eating, excretion, and copulation, the obsessions with genitals and orifices, especially with the president's spectacular penis and anus, allow everyone to participate in an extravagant yet exhausting theater of postcolonial virility. "The very display of grandeur and prestige always entails an aspect of vulgarity and the baroque," Mbembe writes. The autocrat enacts conviviality; the tyrant appears intimate with his subjects. Indeed, "the presidential anus is brought down to earth; it becomes nothing more than a common-or-garden arse [*un anus bien du terroir*] that defecates like anyone else's."[24] Mbembe thus conjures Gargantua and Pantagruel in Africa: "a world hostile to continence, frugality, sobriety," "a world of narcissistic self-gratification," where "the act of exercising command cannot be separated from the production of licentiousness."[25] As Judith Butler observes, "this is simulation that both ridicules and reinstates the commanding power." But we never learn how "the vocabulary of the *commandement*'s self-production and self-aggrandizement itself replays, recaptures, reinscribes the simulations of racialized sexuality commanded by colonial impositions of power" – that is, the extent to which these postcolonial tropes are just dirty colonial leftovers.[26]

From the 1960s, African literature has overflowed with critical excremental signifiers. Novelists turned to shit as a marker of political misdeeds, wasteful over-consumption, and failed development – more often echoing Swift's misanthropy than Rabelais's conviviality. Rather than a fun mode of postcolonial governance, excremental debauchery and anal jollification represent in these texts the "cloacal forms" and peristaltic perversity of neo-colonial capitalism, the corruption of elites, and the miscarriage of the national project. "In the colonial era, shit often functioned as a sign of the actively denigrated native," literary critic Joshua D. Esty tells us, but "it also comes to function, in the decolonization era, as a sign of the actively repudiated ex-colonizer, the alien and unwanted residue of a sometimes violent political expulsion." The divergence from

Mbembe's equally plausible analysis indicates "the dangerous mobility of the excremental signifier in colonial settings" – one might say its promiscuity.[27]

"Huge turds floated in decomposing rings" down streets of rundown Lagos, Nigeria, in Wole Soyinka's *The Interpreters* (1965). The young journalist Sagoe, the novel's principal character, sees himself as a modern hygienic subject surrounded by the shit of underdevelopment. Filth is everywhere; the immeasurable turds that beset him are colonial floaters that cannot be flushed away. "Next to death, he decided, shit is the most vernacular atmosphere of our beloved country." Fearing himself colon-ized, Sagoe struggles to develop a philosophy of shit, a theory of voidancy, thereby re-imaging defecation as a private act of self-strengthening. But his recurrent retreat to the latrine cannot immunize him against colonial pollution. Mimicry of the colonizer is no escape from debasement.[28] As Ghanaian novelist Ayi Kwei Armah put it in *The Beautiful Ones Are Not Yet Born* (1969): Kroomson, a corrupt government minister was "trying to speak like a white man, and the sound that came out of his mouth reminded the listener of a constipated man, straining in his first minute on top of a lavatory seat."[29] The colon-izing of the "native" body thus proves a remarkably durable formation, a lastingly potent mechanism for deferral of civilizational accreditation.

Again and again, we witness the weird afterlives of colonial distinctions between asceticism and defilement, purity and danger, closed orifices and open apertures – a structural truism or imaginary, with variously valued polarities. We seem fixated on analysis of defecation and toilet training, whether in critical or instrumental or celebratory modes. While attempts to valorize excrementality as a valid form of governmentality, or alternatively to read shittiness simply as an index of national failure, might appear unruly and oppositional, they often shrink down to the standard colonial search for a fecal signature. Every de-sublimation begins to look like debasement, the naïve replication of colonial structures of feeling. Categories that appear so excitingly assailable are found to be distressingly durable, resistant to de-colon-ization. Every supposedly disruptive and disturbing return of the repressed, any revelry in anal eroticism, ends in complicity with modern binary classifications. Thus, in trying to escape the structures of modernity, we may find ourselves mired in what Edward W. Said called "the nativist impasse," grotesquely disfigured and reductively libeled, even if self-libeled.[30]

Anthropological Excremental Romances?

While colonial public health officers and other white bureaucrats in the early twentieth century labored, as we have seen, to stereotype subject peoples as promiscuous defecators – ascribing to them a dangerous filthiness, a corporeal depravity so threatening to allegedly pure and vulnerable European colonizers – at the same time an emerging cohort of social anthropologists was becoming enthralled by native scatology. Supposedly primitive customs and habits relating to the anus and feces seemed to offer new insight into human cultural development – and into its deficiencies. Anthropologists joined colonial bureaucrats in turning excremental activity into a litmus test for civilizational achievement, allowing precise cultural staging. Among the more notorious pioneering studies was John G. Bourke's *Scatologic Rites of all Nations* (1891), which examined "stercoraceous practices" – most of them medicinal – around the world. A captain in the US army, Bourke had learned about "filth remedies" and other strange excremental rituals from American Indians, the "savage tribes" he fought against on the western plains, eventually writing them up in what he called a repugnant and mysterious ethnology.[31] When the scatological compilation was translated into German, Sigmund Freud contributed the preface, both attracted and repulsed by an evident lack of repression of coprophilic inclinations among primitive peoples, who play insouciantly with their shit, like infants.[32] In his fascination with excretions, Bourke anticipated the interests of a rising generation of social anthropologists. Even Bronislaw Malinowski, some years later, sitting outside his tent near a village in the Trobriand Islands, would cast nervous glances in the direction of defecating islanders, unable to suppress his curiosity – though as a white man he remained remarkably reticent publicly about his own toileting.[33] "The diarrhea of the diary," Dutch medical anthropologist Sjaak van der Geest wisely notes, "turns into constipation at the threshold of civilization."[34]

Intrigued by Malinowski's reports in *Sex and Repression in Savage Society* (1927) of the relaxed attitudes of "primitive" Melanesians toward defecation, among other issues, Freud in the late 1920s urged one of his more devout acolytes, the Hungarian psychoanalyst Géza Róheim, to investigate anal eroticism, and any associated taboos, among Arrernte – those classic representatives of totemism – in the central Australian deserts.[35] Trained in anthropology in Leipzig and Berlin, analyzed by Sándor Ferenczi in Budapest, Róheim was notorious for the "imperiousness and categorical character of his adult personality" and for "his intransigent and well-nigh Talmudic Freudianism."[36] He enjoyed the company of primitive isolates and relished denouncing civilization for the sacrifices it demands. In 1929, camping outside the Lutheran mission at Hermannsburg (Ntaria), Róheim talked in broken English with Aboriginal men about their sex lives and dreams, hoping for a "transference," seeking to learn more generally about "the importance of the infantile period of development, of sexuality, of the interrelations between the individual and society, of dreams or even of the emotional life and ideals of human beings."[37] He encouraged Arrernte to connect their personal dreams, their unconscious, with the "Dreaming," a term that gained currency during the period he conducted fieldwork, meaning the founding dramas of their group.[38] Usually, it was difficult to differentiate individual reveries from mythology and ritual. Róheim concentrated on local versions of the Oedipus complex, focusing on the abreaction to the primal scene, the conflict between son and father, when the totem substitutes for object loss. But the psychoanalytic anthropologist, a friend of Herbert Marcuse, also was impressed by distinctions of character formation in Aboriginal society, consequences of an apparent lack of anal libido. Blithe indifference to bowel control during infancy, according to Róheim, results in intense narcissism, weak superego, homosociality, exaggerated masculinity, and impulsiveness. Aboriginal adults, he believed, therefore are dirty and generous, free of neurosis, unrepressed, displaying polymorphous perversity, with no care for tomorrow and no sense of urgency. That is, these "primitives" show "an absence of the anal sphincter function."[39] Freud was delighted his follower's psychic projections onto Arrernte seemed to have confirmed speculations in his recently published *Civilization and Its Discontents* (1929): that without the trauma and guilt of potty training, the rigid control of the lower bodily stratum, there can be no proper civilization. "What, have these people no anus?" Freud teasingly asked Róheim.[40]

Róheim's studies of primitive character formation or subjectification prefigured some recent anthropological revalorizations of excrement beyond Europe and North America. No surprise that shit has become especially salient in urban African ethnography. Conducting fieldwork in southern Ghana, van der Geest was puzzled by the continuing widespread acceptance of public latrines.[41] He initially wanted to explore their appeal but later admitted that he "failed to practice full participant observation as far as toilets were concerned." On venturing into the local public latrine: "A penetrating stench hit me in the face. The floor was littered with toilet paper. Squatting people stared at me. I felt terribly uncomfortable, walked past them and left

the place." Instead, he located a private toilet in a missionary's
house. "I slowly realized," van der Geest concluded, "that some
of my concerns about privacy, and my perceptions of dirt and
disgust, were different from theirs."[42] American anthropologist
Brenda Chalfin plunged further into Ghanaian excremental
culture. She rejected the European moral drive to sequester
waste, hoping rather to "disinter a realm of political praxis," to
find an African counter-biopolitics, a new "scata-logics of the
post-colony." "The built environments of sewerage and sanitation," she wrote, "offer people an opportunity to wrest a space
for urban existence outside the grasp of political institutions
and elites." Chalfin endorsed Ghanaians' right to shit as they
pleased, thereby transmuting bare life into alternative political possibilities. She criticized the modern state's excremental
disciplines, its standard sanitary scripts, identifying instead
exciting formulations of the African social body in communal defecation, enticing visions of a carnivalesque Ghanaian
"excremental public."[43]

India has become central in appraisals of the oppositional
potential or even liberatory and de-colon-izing possibilities of
unregulated defecation. For Arjun Appadurai, lingering in the
Indian latrine cannot be avoided when "seeking to define what
governance and governability mean." Human waste management is "where every problem of the urban poor arrives at
a single point of extrusion." Appadurai is beguiled by grassroots "toilet festivals" through which Mumbai slum dwellers
perform their competence in public toilet design and construction, thus enacting the "politics of shit" – querying the "poo"
in "poor," as it were. A "spirit of transgression and bawdiness,"
the carnivalesque even, animates these festivals, inverting "the
harmful default condition of civic invisibility that character-
izes the urban poor." In other words, toilet festivals, revealing
alternative vernacular modes of defecation, "are a brilliant
effort to re-situate this private act of humiliation and suffering
as a scene of technical innovation, collective celebration and

carnivalesque play" – to translate abjection into subjectification.[44] Other scholars have contrasted the Indian rural poor's
"communal bond of defecation" to crass middle-class "acceptance of Western modernity and commodification" in cities like
Bangalore.[45]

But some social scientists engage with Hindu cultures of
excrement, the structuring of purity and pollution, on less flat-
tering terms. They condemn the deeply entrenched cultural
norms and taboos that promote open defecation, especially
when affordable pit latrines are available, resulting in the
spread of enteric pathogens that cause widespread sickness
and infant growth retardation. They argue that India is, in
practice, an intensely fecophobic society, displaying urgency
to remove polluting materials from the home, securely depos-
iting them as far away as possible. Accordingly, private pit
latrines are seen as "ritually impure and also polluting," while
open defecation mistakenly appears "a wholesome activity that
is associated with health, strength, and masculine vigour."[46]
Similarly, Assa Doron and Robin Jeffrey demand more anthropological insight into Hindu defecatory regimes, "to avoid the
facile targeting of the poor as deficient citizens, whose latrine
practices are viewed as a 'primitive' source of social disorder
and disease." They attribute the propensity to defecate in the
open not so much to caste sensibilities as to inadequate hygiene
instruction, filthy toilet facilities, and the residual sociability of
shitting in public.[47] And yet, as geographer Sarah Jewitt points
out, "it is important not to romanticize open defecation."[48]
Whether positive or negative, the cumulative effect of these
sociological accounts is to situate the Indian poor under the
sign of shit – even appreciation of Hindu defecatory authentic-
ity defaults inevitably into stigmatization.

I could give further examples of the excremental turn in contemporary anthropology, the ethnological romancing of the
turd, the emergence of shit ethnographies and topographies,
mostly in Africa and South Asia, but also in other dejected parts

of the world and liminal zones[49] – in what paranoid former US president Donald Trump knows as "shithole countries."[50] In making visible an antagonistic subaltern biopolitics, in recognizing other claims to sovereignty over the lower bodily stratum, these sympathetic studies reflect broader challenges to neo-colonial strategies of modernization and development, where social phylogeny appears to recapitulate, and often arrest, the stages of human ontogeny.[51] But such efforts to transform abjection into grassroots biopolitics, to recognize anality as another way of being modern, struggle against our deep-seated categorical discriminations and fears, pitting countertransference against an ingrained sense of disgust. No amount of anthropological scrubbing removes the fecal stain. Even a positive and extenuating re-evaluation of defecatory practices can be read as merely celebrating the picaresque travels of the uncanny stool in what Paul Theroux, following his mentor V.S. Naipaul, once called "The Turd World"[52] – thereby effectively perpetuating colon-ization, rather than liberating anybody from stigma.

I am reminded of scholarly endeavors last century to explain "objectively" how African sexual cultures, supposedly so free and easy, permitted the proliferation and specific patterning of AIDS on the continent. Although well-intentioned, anthropologists imagined a homogeneous population of primitives, depraved and licentious, victims of their monstrous passions: these experts inadvertently recapitulated colonial tropes and white moral panics.[53] Thus, the colonized body became, as Jean Comaroff puts it, "the projection of a self never fully tamable." Contending with the ravages of the AIDS pandemic, anthropologists were identifying "unrecognizable aliens capable of disrupting existing immunities, penetrating once-secure boundaries at a time of deregulated exchange." Comaroff, however, continues to hope that activism may yet stop abjection returning to bite us in the bum. She suggests that the marginalized and excluded lower bodily stratum might someday generate new political subjectivities and different styles of mobilization, what she calls a "counter-biopolitics," a redemptive sociology – and perhaps a release from standard scatalogics.[54]

As a corollary to this growing anthropological passion for assaying the amplitude and aptitude of the excrement of the world's poor, this *nostalgie de merde* or shit salvage operation, I feel the need to address the corresponding indifference to social analysis of incontinence in "civilized" settings, especially among white people. The elimination of the bourgeois lower bodily stratum – or rather, the presumed lack of elimination from it – seems consistent with the usual racialized repressions and denials. Aside from a few passing remarks in the nursing literature, there has been little anthropological analysis of incontinence in medical settings and hospice care.[55] In such modern laboratory-like situations, in places sanitized and sacralized through scientized ritual, feces seem alien and incongruous, requiring swift removal and sterilization. To outside observers, though not to those who work there, the idea that one could conduct in such settings an ethnography of defecation, as one might in Africa and South Asia, would seem strange. When contained bourgeois bodies suffer uncontrolled chronic diarrhea, they gradually exit the category of personhood, becoming sequestered in care institutions and soon dismissed or forgotten.[56] Their abjection is not readily transmuted into a subject position, even an oppositional one; they are not recognized as agents of the carnivalesque or as eligible for inclusion in any counter-biopolitics. There is no excremental romance in hospice care; rather, we see recoil and disgust, efficient dispatch, and oblivion.[57] Such deficiencies in the scope of anthropologies of defecation perhaps are symptomatic of a continuing intellectual investment in idealized modern bodies, safely enclosed and restrained, imperforate and sterilized, distinct from the degraded, yet seemingly alluring, colon-ized elsewhere.

Shit Envy

Although mostly targeting modern white bodies, the elite North Atlantic art world displays similar symptoms and signs to those identified in excremental literature, social theory, and cultural anthropology. For years now I have been watching for "shit art" and its mischievous performances of transgression. While isolated scatological references have long punctuated the traditional canon of European art, interest in ordure and waste did not really proliferate until the twentieth century, manifested in commitments of Dada, Surrealism, and Futurism to dirt and disorder. Modernist artists like Marcel Duchamp – as in his *Chocolate Grinder No. 2* (1914) – wanted to send "messages about the thin and superficial nature of our civilization or that of others and how lurking beneath the veneer is a deep, excremental darkness fond of making it to the surface."[58] In effect, this often meant activating scenes of primitivism, making the "primitive" expressive in modernity as transgression and profanity.[59] Early in the century, Sigmund Freud and his acolytes had speculated on whether art might be a sublimation of anal eroticism, an aesthetic diversion of primitive and infantile tendencies to smear feces on walls.[60] Radical artists thus sought to make visible art's underlying excremental pattern, to de-sublimate it. As art theorist Hal Foster points out, the primordial attribution is fundamentally a "projection of a particular modern subjectivity onto the primitive" – but it is nonetheless a compelling formulation, especially for would-be rascals and rogues in the art world. As Foster asks: "How many avant-gardist gestures have invoked dirt or shit?"[61]

The wild profusion of shit art during the past sixty years defeats any simple summary. As critic Gerald Silk assures us, there is "no dearth of examples of stercoraceous subjects in the history of art." Predictably, most recent excrementalists – *soi-disant* pranksters and contrarians – have been white men, though a few non-whites and women have managed to slip into the shit-stream.[62] Italian Piero Manzoni is among the more notorious father figures. Influenced by Duchamp, he decided in 1961 to market his own shit in a project called *Merda d'artista*, producing ninety cylindrical cans, each allegedly containing 30g of excrement, each signed by the artist. The cans were sold by weight, their value based on the price of gold at the time – since then they have appreciated exponentially, outperforming gold by a factor of seventy or more.[63] *Merda d'artista* cleverly conflated capitalist production and waste; it caused viewers and collectors to question the status of the art object; it played with notions of artistic commodification, drawing attention to the fetish. Presenting cans of excrement as art was innovative and risky in the 1960s, but whether *Merda d'artista* constituted aesthetic transgression and critique or whether it celebrated emerging consumer culture in late capitalism has been difficult to determine.

Visitors to a mirrored room at the Museum of Old and New Art (MONA) in Hobart, Tasmania, are assailed by the stench of Wim Delvoye's *Cloaca professional* (2009), one of his vast digestive machines constantly excreting shit, or something like it. A Belgian conceptual artist who made his name tattooing dead chickens and live pigs, Delvoye constructed his first defecation apparatus in 2000, after consulting with gastroenterologists and plumbers, hoping to transform food into feces. Nutriment passes peristaltically through digestive or assembly stations before being extruded through the back passage, the rectum, where the smelly waste is collected and put in jars of resin, deliberately not signed.[64] Controlled from a laptop, the "cloaca" disconcertingly combines digestive tract, laboratory, and lavatory, thereby confusing humanity, experiment, and machine. "From the Second World War on," Delvoye said, "I see only scatological art, even when it's not meant to be." The philosophical writings of Peter Sloterdijk along with extensive travels in Asia have informed the artist's excremental vision. When learning woodcarving in Sumatra, Indonesia, Delvoye developed a special rapport with local inhabitants: "There was something in them that made me think they were like me," he recalled – note, *something in them*. He discovered there how to combine high and low: "It's like you see through the anus to see the stars, and you see the stars only through the anus."[65] He was mobilizing shit in part to parody the modern culture industry. "If you make use of scatology, you're not suspected of being very serious, so it gives you a lot of freedom to make fun of things: how products are marketed, how publicity works, how consumers behave, how brands are built, how factories try to sell products that we don't need. . . ."[66] In Delvoye's work, primal waste accumulation has turned into Marx's commodity fetishism.

The media of Latino artist Andres Serrano, based in New York City, are blood, urine, breast milk, semen, and feces – perhaps his most famous composition is *Piss Christ* (1987), a photograph of a plastic crucifix submerged in the artist's urine. In 2008, he took sixty-six elaborately staged photographs of piles of human and animal dung, starting with his own, which he called *Self-Portrait Shit*. "Just before I started to make these pictures," Serrano confided to Lynn Yaeger, "I had a moment of panic: What if I can't find beauty, diversity? What if they don't look good?" But seeing the image of his own shit cheered him immensely. Supposedly, the idea for the series had come while watching the naked wrestling scene in the film *Borat* (2006) – somewhat incongruously, as other scenes might be more pertinent. "I had to prepare myself mentally," he recalled, "it was a scientific and aesthetic investigation." The smell often overwhelmed him, causing him to wear a mask. The lighting was tricky. "I would look at it from all angles." But it seems transgression was not uppermost in his mind. "Many people are quite surprised at the beauty and seductive quality of the work," he said. "After a while it just becomes abstractions." For viewers it was "a very clean experience – just an aesthetic pleasure." Disarmingly, Serrano concluded: "My ego as an artist says I can make anything look good, even shit."[67] Of course, it is not unusual for excremental critique thus to be packed into a narcissistic vehicle.

As anxieties about pollution and contamination gained force at the end of the last century, several politicians and erstwhile defenders of family values took to condemning the imagined transgressions of shit art. When Afro-Caribbean British artist Chris Ofili exhibited his painting *The Holy Virgin Mary* (1996) at the Brooklyn Museum of Art, Rudolph Giuliani, the mayor of New York City, feigned indignation. Giuliani expressed his horror in hearing of the painting, which depicted a Black Madonna surrounded by images from blaxploitation movies and cut-outs of female genitalia from pornographic magazines, spattered with elephant dung. He claimed such profanation of the Virgin was sick. "I would ask people to step back and think about civilization," the famously vulgar and flatulent mayor opined. "Civilization has been about trying to find the right place to put excrement." He continued: "The advance that we had in our civilization was that we figured out how to deal with human excrement, without putting it on walls."[68] The Brooklyn Museum was unfazed. In 2015, Christie's auction house in London sold the painting for nearly three million British pounds.

Despite Giuliani's protests, performances of fecal transgression in modern art, those diverse mobilizations of shit to enact a primitivist counter-narrative, often seem to execute little more than decorative or ornamental functions in a pristine, laboratory-like museum, where they are contained and authorized. But then, artistic renewal rarely correlates with political transformation. In most shit art, there is an underlying implication that the site of excremental alterity lies elsewhere, among the colon-ized, and that a daring association of it with white masculinity or with religious figures will somehow be profane and destabilizing. The work of art thus can be seen as representing shit envy, as depicting the romance of the turd, as just another *"primitivist fantasy,"* suggesting lewdly that "the other has access to primal psychic and social processes from which the white (petit) bourgeois subject is blocked."[69] Shit art therefore is not really a matter of the abject troubling subjectivity; rather, we get a sense of objects taken from elsewhere, appropriations of "authentic" excrement, sealed in cans and jars or alienated in images, shit-substitutes distanced and re-fetishized in museums and art markets. Although so much of the intense focus on embodiment during the past three decades appears to have been responding to the AIDS pandemic, there is nonetheless a feeling of instrumentalism and sustained normativity about this enterprise, a failure of critical engagement. All those melancholy turds offered up safely for exhibition and profit, perhaps to deliver a momentary anal *frisson* among collectors. Such is the avant-garde, so concerned to guard the rear end. Where is the trauma? Where is the abject? Where is the real shit?

When I view shit art, I keep thinking of an episode of *South Park* from 1997, "Mr. Hankey the Christmas Poo." It must be everyone's favorite episode. Kyle, a Jewish boy who wants to participate in Christmas festivities and commerce, to participate in American consumer society, imagines a jolly Santa-hatted turd, Mr. Hankey, who leads him around, singing, dancing, and saying "Howdy-Ho" – as well as smearing everything with shit. Some commentators interpret the episode as an exploration of young men's "expulsive anality" and ambivalent relations to contemporary cultures of consumption;[70] while others regard it as sensual scatology, celebrating grotesque bodies and ecstatic abundance, profane expenditure and monstrous excess.[71] Certainly, Mr. Hankey contributes mightily to making a spectacle of shit. But it seems no one believes this twentieth-century version of Rabelais' *Gargantua and Pantagruel* is particularly transgressive and disturbing.

Dirty Protests

I have followed here a rather circuitous and difficult passage, only to arrive at a predictable destination. It may be that along the way I have come to resemble Jonathan Swift's despised "critick" in *A Tale of a Tub* (1704): someone who "divides every beauty of matter or of style from the corruption that apes it." I feel I may have circumnavigated the romantic recuperation of shit "with the caution of a man that walks thro' Edenborough [Edinburgh] streets in a morning, who is indeed as careful as he can to watch diligently and spy out the filth in his way, not that he is curious to observe the colour and complexion of the ordure or take its dimensions, much less to be paddling in or tasting it, but only with a design to come out as cleanly as he may."[72] Above all, I too have tried to show how we might come out clean.

In this chapter, I have traced multiple endeavors to revalorize excrement, to activate socially and politically the dirty and degrading polarity of a conventional structural dichotomy between purity and danger, between good bourgeois character and anal eroticism, between immunity and infection. Earlier, I described the racial disparagement of the colon-ized – but here I consider what happens when anality and shit, monstrous bodies and the carnivalesque, are mobilized as potential counter-biopolitics, whether in literature, social theory, cultural anthropology, or contemporary art. As it happens, by "mobilized" I really mean absorbed into a socially acceptable matrix. In the end, I find I have been delineating two complementary modes of securitizing and recuperating shit – of coopting it into dominant scatologic discourses, re-fetishizing it, making a spectacle of it.[73] There is the classical style, located in biomedical sciences and epidemiology, an excremental appropriation manifested most effectively in colonial public health, wastewater surveillance, and, as we shall see, in studies of the human gut microbiome. Then there is, as described here, the romantic style, a sort of *détournement* or hijacking of base matter, an ultimately forlorn effort to turn the filthy and profane against the ascetic powers that be. Ostensibly different, both styles result in transforming the abject into a profusion of objects, to which we readily adjust, accommodating ourselves. Despite different goals, both styles are apotropaic, working as charms to ward off evil. The repeated failure to make shit radically corrosive, to generate an alternative biopolitics of abjection, leaves us recursively mouthing the same platitudes, repeating old binary discriminations, stuck in the same optimistic structuralist loop, like a character in a Samuel Beckett play, *Happy Days* (1961) perhaps.

To reveal complicities of literary and artistic excremental representations with structures of power is not to deny that shit might still serve as a weapon of the weak.[74] One of the more prominent examples of scatological entropy and violent resistance to classification occurred in Northern Ireland prisons in the late 1970s during the Troubles. In the so-called Dirty Protest, inmates refused for years to leave their cells to wash or use the toilets, shitting in the corner of the cell and smearing excreta on the walls.[75] As Begoña Aretxaga observed, "the Dirty Protest provoked an inexpressible horror." Any contact with prisoners was abhorrent to the guards; the public, even sympathetic Irish Republicans, recoiled at the defilement. Aretxaga claimed the Dirty Protest was a "materialization of the buried 'shit' of British colonization, a de-metaphorization of the 'savage, dirty Irish.'" Prisoners were "transforming the closed universe of the prison into an overflowing cloaca, exposing in the process a Boschian vision of the world, a scathing critique of Britain and, by association, of civilization."[76] But what strikes me is the *inexpressibility* of the horror: the impossibility of putting into language the destruction of categories that was taking place, the futility of attempting to recuperate the abject as a stable object, to perform any sublimation, to secure any ontology or recognizable theme, to find meaning in such unsublatable excess.[77] The Dirty Protest left everyone speechless, it was so exorbitant a declassification. As Vicente Rafael writes of similar protests in the Philippines under dictator Ferdinand Marcos: "the guards risked being swallowed up into what essentially had become the extension of the prisoners' anal cavities."[78]

5

Gut Feelings and Dark Continents

Sometime in or about the year 2001, ordinary shit-in-the-making became the "gut microbiome." From that time on, countless microbiologists and data scientists have come together to record and to speculate on the interactions between the teeming microbes occupying human intestines, hoping to understand their combined effects on the host body's immune system, metabolism, and mentality.[1] There is certainly a lot to contemplate, with the cells and genomes of gut microbiota, or intestinal flora and fauna, outnumbering the body's own cells and genes many times over. As in wastewater epidemiology, so too in the clinical domain has abundantly animated stool, or proto-stool, become information rich, heavy with data. The microbial multitudes that help to compose our fecal load have been reappraised, turned into a heterogeneous object possessing some as yet undetermined biological value. Intestinal contents, once abject, are now securely object; base matter, now lively matter, verging on carnivalesque. Or so it may seem. Some historians and philosophers of science have focused on altered perceptions of embodiment and biological individuality, the implied shift from a bounded singular human to a complex assemblage of millions of species, a mutually dependent collective, a superorganism or holobiont. Others more medically inclined observe the abandonment of older models of human immunity predicated on a fixed dialectic between self and other, the idea of a body fortified against the foreign, in favor of a relational, co-constituted identity, a dynamic and interactive multispecies immunitary system – a kind of co-immunism. Still others note the switch in attention from pathogenic germs, which must be specially isolated and cultured, toward multitudinous ecological interactions among potentially beneficial microbes, and between these gut buddies and the human body that embraces them – tracing an arc from microbial orthobiosis and antibiosis to probiosis and symbiosis.

These all are important insights. But I am more interested here in efforts to reify the neighborly microbiome, attempts to euphemize and securitize the making of shit – a word never to be uttered – that is, to molecularize and fetishize and safely mobilize human excrement, yet again. While so many scientists and environmental epidemiologists try to explain how the gut microbiome modulates immune systems, they fail to see that re-conceiving shit as the "gut microbiome" is itself a move to immunize against the abject, to cope with forestalled excretion, to instate order on threatening disorder – that is, to displace or ward off perceived danger lurking in our inside outsides. The new microbiome thus inscribes an exonerative paper trail for shit. Seduced by expressive flourishes in the microbial transcript, scientists and humanities scholars alike keep asking what the microbiome says, not what the concept may suppress or silence or redeem.

"Microbes are good to think with," writes multispecies ethnographer Stefan Helmreich.[2] No doubt true, but they are also good to think against and to think beside or around. Anthropologist Heather Paxson proposes a "microbiopolitics," which would examine critically "the creation of categories of microscopic biological agents; the anthropocentric evaluation of such agents; and the elaboration of appropriate human behaviors vis-à-vis microorganisms engaged in infection, inoculation, and digestion."[3] Together, Paxson and Helmreich extol the emergence early this century of microbes "as models and media for transformative food politics, new ecological and biomedical futures, and even studies of life as it might exist in other worlds." They revel in the heterogeneous microbial field, sharing an ecological vision where everything is relational, all boundaries can be breached, the autonomous self has dissolved. Germs have become germinal. The microbial turn reveals "new potentialities latent in organic, biotic nature."[4] Other humanities scholars have joined in praising the scientific recognition of friendly microbial agency, claiming that "the holistic perspective offered by the concept of the microbiome questions the predominant antagonism between humans and microbes fundamentally."[5] That may be so, but the recent microbial colonization of our life worlds, of our conceptual landscapes, like all imperial enterprises can obscure other processes at work, denying longstanding fears and repressions – using our microscopic ecological mates to help suppress our deep anxieties. Therefore, this chapter concerns neither microbial liveliness nor digestion nor defecation, but "fecation" and its disavowal.

After the development of high-throughput, low-cost gene sequencing techniques in the 1990s – frequently making laborious bacterial culturing methods redundant or at best secondary – the rush to analyze our internal microbial companions commenced. It became easy to detect and to sequence the genomes of multitudes of germs in their intestinal habitats, their anthropogenic niches. While remaining wary of *shit*, seeking to imagine it otherwise, researchers could acquire feelings for the *microbiome*, its distillation. At the turn of the century, Nobel laureate Joshua Lederberg, who had discovered mechanisms of bacterial "sex" or conjugation, suggested the term microbiome "to signify the ecological community of commensal, symbiotic, and pathogenic microorganisms that literally share our body space and have been all but ignored as determinants of health and disease."[6] He wondered at the time if the coinage would catch on. Since then, a plethora of studies has charted the dynamic composition of the gut microbiota, usually emphasizing environmental and nutritional influences, sometimes imputing human genetic pressures, invoking gender, race, and geographical origin of hosts. The extent of fecal intelligence, of intestinal surveillance, widened as never before. So far, however, the surge in microbiome research has resulted in few precise explanations of the positive impact of our networked microbial residents on immune function, metabolism, and cognition. Certainly, there is a paucity of validated therapeutic breakthroughs. The intrigue of the microbiome, like the obsession with sewage surveillance, still exceeds any practical justification; its symbolic import outweighs any current medical utility. To be sure, the microbiome is treasured as shit now safely transmuted, having become alienated and objectified and transactional – but its clinical efficacy remains mostly promissory.

This chapter elaborates on the broader theme of how human waste vacillates in value. Previously, I discussed the epidemiological or informational reappraisal of shit entailed in wastewater surveillance; the racialized governmental value discerned in excrement of the colon-ized, which was generatively disparaged in tropical public health; the recurrent romancing of the turd in carnivalesque literature, critical theory, cultural anthropology, and shit art. Here I would like to add analysis of the gut microbiome to the growing list of apotropaic genres, to the litany of ambiguous and unavailing efforts to refigure and reify the abject, to signify away whatever matter out of place we cannot remove and extinguish. Shit can be both compelling and repulsive, secretly desirable yet resisting our desire, something that must be refigured as an object and distanced from us, thereby having some value assigned – a process Georg Simmel called "the metaphysical sublimation of value," a reflection of the "measuring, weighing and calculated exactness of modern times." All the same, we come to imagine such scientifically and aesthetically fetishized waste as possessing *inherent* value, ready to be realized through transactions of new commodities, such as data, stories, images – and in the case of microbiome research, through retention enemas, fecal transplants, and standardized stool capsules, which come at a definite price. And yet, as Simmel put it, "every valuation is supported by an elaborate complex of feelings which are always in a process of flux, adjustment and change."[7] Thus, the human gut microbiome is a further instalment of the usual tale of metaphysical pathos.

Toxic Hauntings in Constipated Modernity

Before getting to the "ecological" microbiome, let me start with autointoxication, its toxicological precursor in biomedicine. Autointoxication emerged to explain clinical problems of the early twentieth century, as a theory postulating that toxins or ptomaines produced by intestinal microbial flora might enter the body, thus inciting degeneration and neurasthenia.[8] According to a leading proponent, Élie Metchnikoff (Ilya Ilyich Metchnikov), awarded the Nobel Prize in Physiology or Medicine in 1908, the colon is an atavistic or zombie organ in the modern body, abounding in primitive bacteria – making us moderns, in a sense, internally colon-ized. With civilization and rigid toilet training, feces are arrested, we become constipated, allowing putrefaction and the build-up of bacterial toxins, which the vulnerable body soon absorbs. The dialectic between modern and primitive does not end well. Metchnikoff proposed to mitigate its dolorous effects by adding beneficial bacteria derived from yogurt to the diet; some experts advocated rigorous colonic irrigation and enemas; while others insisted on surgically removing large sections of the bowel, sometimes with fatal consequences.

At the beginning of the twentieth century, French anarchist and journalist Émile Gautier, inventor of the term "social Darwinism," told readers of a popular journal that "the danger zone of our organism, the real defect of our armature, is the intestine." The digestive tract is the body's sewer, awash with food residues and multiple microbial species, assembling themselves into feces, "the seat of abominable putrid fermentations." Modern life was causing increased intestinal stasis, promoting the ingress of accumulating toxins through the gut's permeable walls. "There's something rotten in the lower regions," he warned. Immunologists at the Institut Pasteur in Paris had demonstrated that most human diseases derive from autointoxication, the result of bacterial poisoning. Gautier expressed support for their plans to introduce a digestive "police force" consisting of replacement "good microbes . . . capable of ceaselessly opposing fresh troops against the swarming enemy." He, too, recommended the powerful lactic ferments of Bulgarian yogurt. One day, he speculated, these beneficial germs and their secretions might be put into tablets or capsules, preforming "the double task of drainer [*vidanger*] and gendarme" of our bowels.[9]

Evidently, Gautier was heeding the dietary advice of Metchnikoff, a Russian embryologist turned immunologist, based at the Institut Pasteur, famous for his discovery of phagocytic cells that roam the body, engulfing and consuming foreign substances, including bacterial fragments.[10] The first line of natural immunity, phagocytes (later macrophages) appeared vulnerable to senescence and impotence, affected by toxins emanated from putrefactive bacteria in the colon. According to Metchnikoff, the colon is a kind of vestigial cesspool, blocked up by the detritus of Western diets, rendered noxious by advances in civilization. A primitive organ, it allows humans to store waste until it is safe to excrete, but civilization, which should which should in practice permit defecation without peril, renders retentive bowels both unnecessary and harmful.[11] Metchnikoff was intrigued by recent studies conducted by Charles Bouchard, a pathologist who had worked (like Sigmund Freud) with neurologist and expert in psychological diseases Jean-Martin Charcot at the Pitié-Salpêtrière Hospital in Paris. In the 1880s, Bouchard had become convinced that the stasis of "lower organisms" in the human digestive tract causes internal poisoning, which he described as "autointoxication."[12] Of course, Bouchard was following a long line of French pathologists eager to ascribe all diseases, especially mental ailments, to gastrointestinal disorders and visceral nuisances, to a lack of sympathy between innards and the rest of the body[13] – only now the malcontents were microscopically animated.

Metchnikoff's concerns with corporeal integrity and defense had evolved through colonial experiences on the central Asian steppe. There he became aware of the danger of the primitive dwelling within the modern corpus. Engrossed in the ethnological study of Kalmycks, the Mongolian Buddhist inhabitants of the steppe along the western shore of the Caspian Sea, Metchnikoff came to believe these "primitives" were still physiologically harmonious, unlike constipated civilized men like himself, who contained the contradictions of modernity and barbarism.[14] Organs like the colon, atavistic in civilized bodies, seemed to present no threat to non-modern Kalmycks. Fascinated, Metchnikoff spent the rest of his life endeavoring to work out how to harmonize the incongruities of mixed-up modern bodies, through harnessing primal elements like phagocytes, hoping thus to extend the longevity of white men. He imagined phagocytes as African imperial conscripts policing the body's boundaries, holding the truly wild gut microbes at bay. Primeval themselves, these "white" blood cells could be drafted and trained to control the true colonic savages. But Metchnikoff also realized that the barbarism within us could never be fully eliminated – a depressing thought that prompted one of his many suicide attempts. Moreover, the interests of phagocytes did not always correspond to the overall needs of sovereign white selfhood. The immunologist witnessed the primitive recruits turning on the superior cells of the body, rebelling against them, gobbling them up. "It is a veritable battle that rages in the innermost recesses of our being," he wrote.[15] Perhaps, he wondered, damaging toxins from savage gut bacteria were weakening or confusing phagocytes, causing them to attack "noble" cells by mistake. Therefore, he proposed a sort of ethnic cleansing of the bad intestinal microbes – a procedure he called "orthobiosis," or right living, distinct from symbiosis, living together – replacing them with good settler germs like *Lactobacillus bulgaricus*, from reliable Slavic yogurt, with the goal of de-wilding the colon. It would be more like settler colonialism in Algeria than the colonial civilizing process prevailing in places like the Philippines.

Theories of colonial intoxication, as propounded by Metchnikoff, were applauded in Britain early in the twentieth century. Prone to comparing poor humans to compacted fecal material, H.G. Wells, for example, found the immunologist's "exposition of the large intestine as a sort of death-organ in man . . . extraordinarily convincing."[16] In the novelist's *History of Mr Polly* (1910), the eponymous hero's guts are in a state of civil war, resulting in dyspepsia and irritability, conditions later resolved by better diet and residence in a rural idyll.[17] Intestinal self-poisoning also entranced medical luminaries in London. A leading surgeon, and friend of Metchnikoff, W. Arbuthnot Lane, believed that "autointoxication reduces the resisting power of the individual to the entry of organisms of various sorts and facilitates their obtaining a foothold in some of the tissues of the body." Among elite patients he saw a distinct "depreciation in the vitality of the tissues," manifesting as stained and inelastic skin, cold extremities, abdominal pain, headache, vexation, lassitude, and mental distress.[18] In order to alleviate the white race's inner burden, Lane regularly performed extensive resections of the pathogenic large bowel, including all of the cecum, hoping to eliminate any savage races of microbes lying there in wait. Those patients who survived the operation often reported renewed vitality and unaccustomed lightness of step.[19]

American health reformers seized on autointoxication with particular vigor. At Battle Creek Sanitarium, Michigan, John Harvey Kellogg became obsessed with intestinal stasis and bowel sepsis.[20] "Intestinal toxemia or autointoxication is the most universal of all maladies," Kellogg declaimed, "and the source of autointoxication is the colon with its seething mass of putrefying food residues." The errant Seventh Day Adventist physician preached that the colon is "prepared to serve the purpose of a sewer for the body," to carry away putrefying bacteria. "The second half of the colon, the terminal part of the food tube," he wrote, "has no other function than to transmit and eject from the body the waste and poisonous matters that constitute the stool." But modern life was impeding the natural function of the colon, constipating the race. "The whole civilized portion of the human race is housebroken," he lamented. Thus, "the intestinal flora must be changed. This is absolutely necessary to conquer constipation." Kellogg urged on white Americans "the necessity of exchanging the wild bacteria for the protective germs which nature provides as a bulwark against disease." The inventor of corn flakes and peanut butter proposed dietary modification and frequent colonic irrigation to sort out fecal matters and so rescue the white race. "A new and sensitive colon conscience must be developed among civilized people," Kellogg reflected, "if the world is to be saved from the soul-and-body and even race-destroying effects of universal constipation and world-wide autointoxication."[21] Nuts, he often asserted, may save the white race.

Despite the valiant campaigns of Metchnikoff, Lane, Kellogg, and their followers, autointoxication or colonic poisoning was discredited in the 1920s as a threat to white bodily integrity.

Physicians were more likely to attribute common symptoms of distress to mechanical distension and irritation of the bowels – or to psychosomatic problems. Thus, San Francisco physician and medical writer Walter C. Alvarez sought to demonstrate that "in sensitive people the brain is profoundly influenced by afferent impulses coming from a distended, overactive or wrongly acting bowel." He had shown on many occasions that symptoms of so-called autointoxication were produced simply by inserting a cotton tampon or any large object into the rectum. Several of his patients also displayed "colonic phobias" and neurotic or psychopathic personalities. Therefore "we should be slow to accept the enthusiastic claims of rough and ready surgeons who have short-circuited a few colons."[22] A few years later, southern Californian physician Arthur N. Donaldson noted that "the large bowel has, in truth, been the theorist's playground." To settle matters he decided to conduct a trial of constipation on five white men: himself and four medical students. They soon experienced general malaise, feeling rotten and sluggish. Immediately after an enema, each perked up, exhibiting a "decidedly different" demeanor, their mental depression lifted – far too quickly to be explained by elimination of toxins. To confirm his findings, Donaldson then meticulously packed the rectums of his four students with "cotton pledgets soaked in petroleum and dusted with barium." On voiding their rectal load, the young white men suddenly felt like themselves again.[23]

Increasingly, psychoanalytic explanations positing infantile repression and consequent neurosis would prevail in explanations of abdominal discomfort and low mental tone. As Sigmund Freud concocted theories of anal eroticism and the psychological impacts of parent–child conflict over toilet training, he was reading Metchnikoff with considerable interest. In effect, Freud began to displace medical worries about direct fecal toxicity, the trepidation that the rectum may be the white man's grave, onto psychic repression of libidinal aspects of defecation – fashioning an anal seduction theory. Eventually, Freud's arguments for the influence of repressed anal pleasure would substitute for Metchnikoff's colonic materialism – until, that is, the decline of Freudianism late in the century and the emergence of a gut microbiome.

Even so, the "civilized" public continued to fear constipation and to consume what became known, in the 1970s, as probiotics, supposedly beneficial microbes. A few odd microbiologists still insisted on vainly scrutinizing and culturing the bacterial composition of the human intestine, the dark continent within us. Perhaps the most vivid example of persistent captivation by racial peristalsis can be found, not surprisingly, in late-colonial Africa. In the 1960s, Denis K. Burkitt, an Irish epidemiologist who had studied cancer in colonial Uganda, became mesmerized by fecal transit time, believing that intestinal stasis might generate carcinogens. As people "develop," their diets change, losing fiber or roughage, resulting in constipation. All of us are therefore becoming far too modern, squandering primitive élan. Burkitt accumulated epidemiological information on stool size and bowel transit time across the southern hemisphere, comparing "primitive" rural populations with developed urban dwellers, who presumably eat more refined food. He began collaborating with white South African biochemist Alexander R.P. Walker, who also was fascinated by human fecal variation, which he observed in specimens in government laboratories.[24] Walker's research had demonstrated excretory differences between white and African prisoners, with the latter group defecating larger stools more frequently. Together, Burkitt and Walker were able to confirm the colonic stimulus and potential human health benefit of dietary fiber.[25] As Sebastián Gil-Riaño and Sarah E. Tracy point out, they "put crap back on the map" – and in so doing, they told an anti-modernization story that romanticized "primitive" shitting and revalorized the colon-ized.[26] The dietary fiber hypothesis thus contributed to an old medical genre, reproving the pathologies of progress, the drawbacks of civilization.[27] Intestinal stasis with or without toxemia or carcinogenesis represented deeply felt anxiety about the unnaturalness of modern ways of life, a sense of urgency in coming to terms with the unstable ecology of human health – which later concern with the human gut microbiome would sometimes reiterate. Although the new amiable microbiome appears to represent a telling counternarrative to the autointoxicating effete colon, nonetheless the same fetishized colonic other – a residual host–parasite dynamic, a civilized–primitive dialectic – lingers, haunting the ecological frame.

From Toxic Dialectic to Ecologic Harmony, or Not?

Medical absorption in shit – whether concentrated on its toxic or lively or richly textured qualities – never completely faded in the twentieth century. But a great efflorescence of interest in human intestinal microflora, especially in their genomes, occurred in the early years of this century[28] – just as wastewater epidemiology was also taking off, tackling analogous "sewers." The maturation of high-throughput genetic sequencing allowed complete reporting of the human genome by early 2001. Before long, scientists called for sequencing of the human microbiome as well. As Julian Davies, past president of the American Society for Microbiology, claimed: "We depend on more than the activity of some 30,000 genes encoded in the human genomes. Our existence is critically dependent on the presence of upwards of 1000 bacterial species . . . human life depends on an additional two to four million genes, mostly uncharacterized."[29] Davies and others demanded more attention to elucidating genetically the synergistic activities of humans and their obligatory commensals. Microbiologists at Stanford University promoted a second human genome project, this time including the microbiome.[30] At least 233 human genes possessed homologs found only in bacteria; and the human genome also includes multiple endogenous retroviruses.[31] Microbes, the Stanford scientists declared, are thus "defining the very essence of who we are." New gene technologies and bioinformatics might therefore reveal mysteries of both pathogenesis and normal regulatory exchanges between human bodies and their microbiota. "We are still woefully ignorant about the composition and variability of our endogenous microflora . . . It is time to embark on a comprehensive genomic inventory of the large portion of cellular life within the human body that has been ignored so far."[32] We need to understand that the somatic genome is just a small part of the relevant environmental genome, a fragment of the metagenome, the genetic repertoire or genomic continuum of all the diverse organisms occupying an environmental niche that we designate as human being.[33] In these persuasive manifestos, molecules and numbers always substitute for any mention of feces.

Following an international meeting in Paris, which recommended a Human Intestinal Metagenome Initiative, the National Institutes of Health in the United States established in 2007 the Human Microbiome Project, designed to explore four sites: the gastrointestinal tract, mouth, vagina, and skin.[34] It was conceived as a departure from the conventional practice of isolating and culturing a single pathogenic microbe; rather, the large-scale research project would use new genomic technology to explore microbial diversity in the four selected sites, taking samples from several hundred individuals. The plan was to assemble a "standardized data resource" that might eventually show associations with various medical conditions, or at least indicate some connections between the human genome and the microbiome. Shit would be transformed into a massive genomic database. The microbiome project was meant to recruit "normal," not necessarily "healthy," research subjects across the United States, aiming for equal numbers of men and women. "Efforts were made to recruit a sample that was reasonably diverse in terms of race, ethnicity and other demographic features."[35] Predictably, the emphasis on gendered and racialized participant categories often inadvertently gave misleading and deterministic biological credence to cultural assumptions, disguising more complex ecologies, arranging variation simply according to self-assigned "race" and binary gender, matching convenient American classifications. As usual, "normality" and "diversity" would acquire distinctive vernacular inflections.[36]

Within a few years, 242 adults, now deemed "healthy," had been sampled at between fifteen and eighteen body sites up to three times – though the majority of specimens were fecal – generating more than 5,000 microbial taxonomic profiles and allowing the sequencing of more than 800 human-associated reference genomes. As investigators reported hopefully: "The data represents a treasure trove that can be continually mined to identify new organisms, gene functions, and metabolic and regulatory networks as well as correlations between microbial community structure and health and disease."[37] Additionally, the International Human Microbiome Consortium was set up to coordinate research worldwide and to ensure data quality standards are maintained. Not surprisingly, the microbiome rush of the early twenty-first century soon accumulated more data than could be analyzed effectively; the new discipline of bioinformatics struggled to make sense of what was going on in our guts. Despite speculations on "modulation" of mood – "fecal phrenology" as some rakish scientists call it[38] – and subtle inflection of immune function, few plausible causal connections have been discerned, so far, between microbial ecologies and patterns of human health and disease.[39]

The Human Microbiome Project closed in 2013, having stimulated wider university and commercial research interest in the digitalized potential of human intestinal flora. Entrepreneurial scientists in the United States set up the

American Gut Project (AGP; www.americangut.org) in 2012 and sponsored the British Gut Project in 2014, as open-source, community-driven microbiome sequencing endeavors. For a payment of $99, participants across the United States, regarded as "citizen scientists," receive a kit for their fecal samples, which are analyzed and reported back to them. An ambitious, stolid Pakeha New Zealander, Rob Knight, with a Ph.D. in ecology and evolutionary biology from Princeton, led the AGP, described as "the world's largest citizen science microbiome project," which around 2018 was folded into the Microsetta Initiative, based at UCSD (www.microsetta .ucsd.edu). In 2015, Knight had published *Follow Your Gut*, derived from his TED talk. Other gut projects soon cropped up across North America, Western Europe, and East Asia – but not in Africa and the Pacific. In the past few years, Martin Blaser (a campaigner against extinction of the gut organism *Helicobacter pylori*) and Maria Gloria Dominguez Bello (who studies the gut contents of supposedly isolated Yanomami in Venezuela), both at Rutgers, New Jersey, have worked to establish the Microbiota Vault (www.microbiotavault.org), a repository for preservation of human microbiota "for future generations," modelled on the Seed Vault in Svalbard, Norway. Located in Switzerland, the Microbiota Vault – the Noah's Ark of shit – will set protocols and standards for fecal collection and analysis, as well as arranging to freeze diverse samples, thereby saving valuable intestinal germs from the ravages of urbanization and modernization.[40]

After the AGP laboratory sequenced his gut microbiome, food writer Michael Pollan began to think of himself "in the first-person plural – as a superorganism." He realized that his intestinal contents form "a vast interior wilderness that scientists are just beginning to map." Gut microbiota, he was told, "appear to play a critical role in training and modulating our immune system." At dinner with scientists, they discoursed fluently on feces, speaking of their "radical reevaluation of the contents of the human colon." Later Pollan talked with Dominguez Bello about her dedication to accumulating hunter-gatherer excrement. "We want to see how the human microbiota look before antibiotics, before processed food, before modern birth," she told him. "These samples are really gold."[41]

The philosophical significance of the microbiome was immediately alluring. For years, Lynn Margulis had been arguing for the need to recognize symbiotic relationships as refiguring biological individuality, to rethink microbes not as contaminants but as valuable interdependent agents. She inspired others to challenge the notion of the singular individual organism, to blur the boundaries, to incorporate persistent symbionts – thereby constituting a superorganism or "holobiont," an assemblage of the host and its colonial microorganisms.[42] Maybe we can never be totally modern; we are always also microbial. Stefan Helmreich suggested we rename ourselves *Homo microbis*, seeking to realize a "symbiopolitics," respecting the "densely political relations among many entangled living things – not just microbial – at many scales." That is, symbiopolitics would be "the politics of living things coexisting, incorporating, and mixing with one another."[43] Other anthropologists also got caught up in microbiophilia, urging new "microbial humanities." They claimed that our dynamic interactions with microorganisms disrupt conventional definitions of self, whether immunological or cognitive or genetic. We need to think of the microbiome as "co-constituting the metaorganisms we humans are."[44] Perhaps the individual's personal pronouns should be biologically plural. But some fastidious philosophers of biology looked askance at the celebration of such a human superorganism, our relational being. They distrusted glib chatter about self and individuality, suggesting instead that while microbiota might influence or shape human identity, they did not necessarily constitute it.[45] Perhaps not holobionts after all, humans merely carry around microbial communities with which they interact, generally physiologi-cally, sometimes pathologically.

For all the talk of symbiosis and ecological continuity, stud-ies of the human microbiome in practice retain the permeable boundary between the body and its commensals, the old dia-lectics of self and other, the binaries of civilized and primitive – even implications of purity and danger. Far from being integrated into the body, the microbiome usually is reified, represented as a separate and clearly defined object, which occasionally sends out genetic schleppers to cross corporeal borders. Moreover, the molecular sublimation of shit, with its hints of microbial comity, has never quite suppressed the sense of disordered base matter lurking within. Fecal specimens fail to be readily neutralized as lively and pure accessory commu-nities. Indeed, intestinal contents still feel different from body tissues; we have not yet managed to become excrementally post-human. Despite rigorous and intense datafication and securitization, abjection so far has not dwindled into mundane dysbiosis, a vague condition where the symbiotic community falters in diversity and balance, where microbial ecology is out of joint.[46] Our hospitality toward shit still has limits; it remains abject; we would rather it not stick around, but it does.

Rewilding the Colon?

Conjectures about what makes a gut microbiome healthy or good, and what makes it harmful, reveal continuing resistance to thinking ecologically, to disassembling the old structural binaries. Rather than relate the functioning of microbiota to dynamic circumstances, we still tend to assign an absolute value to what are imagined to be their intrinsic or essential qualities. Remarkably, we often map nostalgic racial or civilizational dis-tinctions onto such seemingly autonomous attributes. Again, "primitive" is generally regarded as beneficial while "modern" appears degraded. Thus, "good" gut microbes feast on fiber and natural diets, especially those containing yogurt, kimchi, sauerkraut, kombucha, and kefir, whereas "bad" microbes thrive on processed foods.[47] "It romanticizes our relationships with our microbes," writes science journalist Ed Yong, "painting them as happy partnerships that were better off in the good old days."[48]

From the beginning of human microbiome investigation, scientists expressed concern that their surveys were failing to capture gut microbial diversity in less developed societies. As usual, most of the databases were replete with results from analysis of Western European and North American fecal specimens. Little was known about the geographical and racial distribution of human microbiota.[49] The effects of ethnicity, diet, lifestyle, environment, and patterns of antibiotic use remained obscure. Some brave scientists had set off on expeditions to South American rainforests, Southeast Asian jungles, and African savannahs to collect poo specimens from isolated, relatively untainted tribes. A few microbiologists delved for microbial treasure in the feces of supposedly "uncontacted," yet extensively studied, groups such as the Yanomami, hoping eventually to benefit prosperous white consumers.[50] But the deluge of materials from "developed" research subjects rapidly submerged these studies of excrement on the margins. Scientists therefore called on "stakeholders" to "consider how to ethically prioritize and incentivize improved global representation of microbiome samples." However, they worried that such invigorated global sampling may be directed mostly to improve the health of Europeans and North Americans rather than those who donated out-of-the-way stools. They feared a reversion to "helicopter research" rather than genuine collaborations with national scientists and respectful engagements with local communities – a repeat of "common extractive tropes in imbalanced research projects."[51] Although continuing to ignore the microbial diversity of most of the world's human population surely is untenable, even unethical, scientists need also to be wary of a contrary problem – indulging in what anthropologist Amber Benezra calls unethical "salvage microbiomics." Benezra warns against such efforts "to save valuable, vanishing microbes from modernization without acknowledging the research's own embeddedness in technoscientific systems responsible for changes in microbial populations."[52]

In 2014, Jeff Leach, a self-described anthropologist of defecation, earnestly told Emily Eakin at *The New Yorker*: "We need to go to places where people don't have ready access to antibiotics, where people still drink water from the same sources that zebra, giraffes, and elephants drink from, and who still live outside."[53] An adventurous and entrepreneurial Texan, Leach recently had returned from living with the Hadza in Tanzania's Rift Valley, a tourist hotspot, where he collected stool specimens for Rob Knight's laboratory and the Stanford laboratory of Erica and Justin Sonnenburg, microbiologists disturbed by loss of diversity in San Francisco Bay Area intestines.[54] "Hadza kids are born in the dirt, play in the dirt, and they're literally chewing on animal bones," Leach informed Eakin. "They're covered in microbes, and it's been like that for thousands of years. Maybe because we've unwilded our children, that might play a role in some of the diseases we see in them." Leach was proud that the supposedly uncontacted tribe call him Doctor Mavi, which he said is the Swahili word for shit.[55] In the tradition of self-experimentation, Leach once inserted a turkey baster filled with Hadza feces into his rectum to rewild himself. "I probably had the most diverse ecosystem of any white man in the world," he claimed.[56] On a later expedition, accompanied by an epidemiologist (and nutrition impresario) from King's College, London, and a BBC radio production crew, Leach showed, through his "cherished poo samples," how the local diet and lifestyle could at least transiently increase the diversity of his own "microbe organ," a euphemism for the colon.[57] This time, he coyly desisted from performing the turkey-baster spectacle. As Gina Kolata asked in the *New York Times*: Why are the relatively well-connected and touristed Hadza being "treated as proxies for Paleolithic people"? How could this "research" possibly benefit them? Kolata questioned Leach's qualifications and reported that several women in Terlingua, Texas, had accused him of sexual assault.[58] The microbiologists at Stanford distanced themselves from their collector of primitive shit; and Leach's name was removed from the website of the American Gut Project.

Rumors of the immunological boon of constituents of wild human feces proved unstoppable. Since the 1980s, a few scientists on the edges of immunology had been wondering if modern obsessions with personal hygiene and sterilized environments might render "civilized" immune systems decadent and feeble, raw and inexperienced – therefore prone to allergies, asthma, and autoimmune diseases, which manifest when the body's hyperactive defense mechanisms turn against its own tissues.[59] Early this century, some scientists turned their attention to the recent extirpation of helminths, including hookworms, from the gut microbiota of people living in developed countries.[60] Perhaps the latrinoscene, a century of hookworm eradication efforts led by the Rockefeller Foundation, had over-sensitized or disrupted human immune systems and perversely made us susceptible to internal pathological processes. Maybe improvements in colonic hygiene and toileting, the colonial repression of promiscuous defecation, had drained off some crucial immune-modulating gut microbiomes, leaving us soft and vulnerable. Several researchers saw the need for helminth trials to rewild our colons, arguing for "the selective reintroduction of parasitic worms as 'gut buddies' to tackle autoimmune disease."[61] Eager to test the hypothesis personally, keen to engage in idiosyncratic citizen science, Jasper Lawrence, a British businessman based in Santa Cruz, California, traveled to Cameroon, still a hookworm paradise, to infest himself with nematodes: before long, he was convinced that the supplementary bowel companions had reduced his asthma and hay fever. "Everyone is concerned about biodiversity in the outside world, and saving the rainforest," Lawrence told journalist Tim Adams, "but we've also screwed up the biodiversity inside us."[62] The benefits of infestation stimulated Lawrence to set up a company to breed and sell well-cleansed hookworms, at a considerable price, to consumers desperate to infect themselves. Catering to the market of mostly white middle-class North Americans afflicted with allergies and autoimmune diseases, Lawrence opened the main office in Mexico, since it is illegal to take helminths outside a body north across the border.

In 2012, Californian science writer Moises Velasquez-Manoff ventured to the Tijuana clinic to see if a dose of hookworms would fix his asthma, food allergies, and autoimmune alopecia. "Call it stupidity, idiocy, madness," Velasquez-Manoff wrote. "Call it lunacy. I consider it biological gonzo journalism." Like so many others, he felt as if his "immune system had suddenly and inexplicably gone berserk" – and he wanted a remedy, even a parasitic solution, perhaps a thorough "reconstruction of the superorganism." In a clean spare room in Tijuana, a nurse in a white coat, with gloves, applied a bandage against his left biceps. Velasquez-Manoff soon reported a little burning and itching, the sensation of thirty "microscopic hookworm larvae burrowing through my skin."[63] Later, he claimed that his new gut buddy, *Necator americanus*, may have helped with some symptoms, but not so much that he would welcome it again. Others have emerged from the hookworm underground more enthusiastic about potential immune modulation. William Parker, a biochemist associated with Duke University, believes colonic hookworm infestation can improve allergies, mental health, migraine, autism, and autoimmune conditions like multiple sclerosis. He condemns the Rockefeller-led "genocidal campaign" against intestinal parasites, which resulted in the "slaughter of billions of innocent and even helpful worms," shamefully stigmatizing them as germs of laziness. Parker has dedicated his life to restoring gut health, combating biome depletion, using helminthic therapy to rewild the colon. He fears that pharmaceutical companies are trying to suppress biomic warriors, guerilla poo fighters like him, given that it is exceedingly hard for big corporations to patent and make shitloads of money out of shit.[64]

Interest in fecal transplantation or grafting has burgeoned in the past twenty years.[65] But the most celebrated precedent, and, so far, the only reliably successful intervention, took place more than forty years before the word "microbiome" became popular. In the late 1950s, Denver surgeon Ben Eiseman was seeing ever more patients with profuse diarrhea, often fatal, resulting from overgrowth of *Clostridium difficile*, a bacterium resistant to common broad-spectrum antibiotics. A former naval surgeon and tropical diseases expert, Eiseman daringly decided to treat four cases, three of them white men, with retention enemas "composed of normal feces suspended in saline," donated by pregnant women, likely to be healthy. It was a simple procedure to correct a presumed deficiency. The results were astounding, with the introduced microbial flora outcompeting the pathogenic *C. difficile* and recolonizing the gut.[66] To this day, fecal transplantation, rationalized by a specific Darwinian process, is the standard remedy for the condition. With the flaunting of charismatic gut microbiota in the past two decades, clinicians have tried to treat other intestinal disorders, such as Crohn's disease, ulcerative colitis, and irritable bowel syndrome, with similar transfers of feces, but the outcomes of such trials have been relatively modest – any slight improvement is often evanescent. Fecal transplantation is a crude form of bacteriotherapy, an unrefined and unprocessed probiosis, delivered through endoscope, naso-intestinal tube, retention enema or non-chewable capsule (often referred to as a "crapsule"). Donors are screened to rule out any pathogen carriage or disease and their stool is assessed for microbial diversity. Increasingly, some fecal sources are regarded as super-donors, especially fecund in good bacteria, excreting an unusual compelling microbial signature. The strongest indicator of transplantation success is "the ability of the donor to transfer high levels of particular keystone species to recipients," in contrast to the old "one stool fits all" approach.[67] Nonetheless, we still have a long way to go before achieving the ideal of bespoke super-shit.

As Emily Eakin recognized a decade ago: "Biotech companies are competing to put stool-based therapies through clinical trials and onto the market. In medicine, at any rate, human excrement has become a precious commodity."[68] No longer is it the usual old shit – but neither is there much evidence yet for the preventive or therapeutic value of such charmed microbiota. Nonetheless, widespread fascination with the potential benefits of animated feces may overcome or suppress what is longstanding repugnance. To meet rising demand, several MIT graduate students in 2013 established OpenBiome, a stool bank that soon was shipping frozen "standardized" poo to most parts of the United States.[69] When Eakin visited the laboratory in Cambridge, Massachusetts – down the road from where William James had opined on matter out of place and Amy Rockefeller unveiled her self-composting latrine, not far from the manhole where C.-E.A. Winslow began his wastewater surveillance – she observed a technician "preparing for the day's stool donations by donning protective gear: a white coat, safety goggles, surgical gloves." She learned that all prospective donors, many of them local students, go through a lengthy screening process, involving blood and stool tests – and swearing they have not recently traveled to the "developing world." Any hunter-gatherers from Africa and Amazonia passing by would be deemed ineligible, far too risky. In any case, more than 97 percent of local homebody applicants still are rejected. The admitted donors – mostly white males bearing fetching nicknames like Winnie the Poo, Allbutt Einstein, and Vladimir

Pootin – are expected to bring in their turds still warm and steaming. They receive $40 for each specimen and become eligible for the monthly prize for the weightiest contribution. The technicians then mechanically separate the slurry of microbes from "food wastes" in each container, transferring the valuable microbiota via pipette to sterilized plastic bottles, which are frozen, ready for distribution.[70] In the past few years, scores of other stool banks have followed OpenBiome's lead, making commercial quantities of safely processed shit – avoiding as much as possible the yuck factor – available across the developed world. Meanwhile, the Food and Drug Administration ponders whether donors should be anonymous or known to the recipient, and whether feces should be classified as a drug or as a tissue, like blood and bone grafts.

Frustrated by the lack of widespread institutional and professional support for fecal transplantation, "citizen scientists" have taken to the internet to arrange their own experimental exchanges. Thus, citizen science increasingly sediments as citizen excremental salience. Tracy in Melbourne, Australia, set up a popular website, The Power of Poop, as an information resource for "e-patients" who hope to negotiate their transplants, as a means to enhance microbiomic prominence and to link donors and recipients.[71] Her "team poop" consists of "normal people living in all corners of the globe, who have been on a digestive illness journey and discovered that we have a microbiome: a microbial miracle within that left alone will nourish, nurture and protect us – until we mess it up." Fecal transplants, the authors insist, are the ultimate probiotics, able to recolonize the colon, constituting the perfect cures for intestinal dysbiosis. But they ask, "why is it so hard to get treatment? Why is it seen as a last resort? Why are risky antibiotics, immune-suppressants, anti-inflammatories, and anti-depressants seen as better options? ... It's time to get down and dirty and talk openly about poop." In The Power of Poop, Tracy and others share online their journeys as e-patients in search of feces. A middle-class white woman in her fifties, Tracy suffered from migraines, anxiety, depression, brain fog, "derealization," and food intolerances, which she associated with escalating antibiotic use. After defecation, her mood would lighten. She observed that her excreta appeared to be "fermenting and smelled sour like yoghurt or wine." Scouring the internet, she learned about Thomas Borody, a controversial Polish-Australian expert in tropical medicine who had founded the Centre for Digestive Diseases in Sydney in 1984, to promote fecal transplantation. By 2020, he could boast of having infused microbially diverse feces into thousands of depleted colons – before getting distracted around then, pivoting to promotion of ivermectin as a cure for COVID-19. Tracy rushed to Sydney in 2011, overjoyed that Borody might cure her gut dysbiosis. After repeated transplants, many of them later conducted at home with a donor she befriended, Tracy emerged healthy from "a vicious cycle of dysbiosis, intestinal permeability, immune disruption and systemic chaos." Just to be sure of remission, she also embarked on helminthic therapy, which seemed to reduce any residual gluten intolerance.

Like many other fecal seekers, Tracy insisted on finding a healthy white male donor. "I approached a friend," she told her followers, "who is a healthy athletic type who never gets travelers' diarrhea when travelling in Asia." She feared humiliation, but Stephen was willing, in his words, "to open his mind to this bizarre request." Tests "confirmed a robust and balanced microbiome." Stephen believed that "with my record of almost perfect health and a self-proclaimed 'stomach of steel' I'd be an ideal candidate to donate." Nonetheless, it took "a complete paradigm shift to embrace poop." After being "groomed" by Tracy, he soon came around, regularly racing across the city to deliver fresh stool for her retention enemas, happy to assist his friend in her fecal journey. Since becoming a donor, he reported, his shit's "aroma is less unpleasant."

When I read Tracy's story and corresponding tales from the microbiome underground, I see continuities between emerging modes of fecal transfer and older, more established shit art – and virtually no connections with formal ecological reasoning. That is, I see Tracy and her kin as combining popular microbiology with eco-art, or environmental art, and shit art in novel ways. Fecal transplantation entrepreneurs and more conventional shit artists seem to possess in common a fascination with stool as aesthetic object and commodity, an excitement with its vacillation between filth and beauty, waste and valuable, its instability of form, threatening disorder. There is a sort of convergence of the medical and commercial mobility of feces with shit as mobile aesthetic signifier, except one claims the imprimatur of science and the other flaunts the allure of critique. Perhaps the emerging science of the gut microbiome is really just shit art all the way down – which is meant as a compliment, of course.

From Bare Life to Bioavailable Shit?

At this point, attempting to wrap up or wipe up the soiled contents of this chapter, I might focus on the dissimilarities between shit configured as autointoxication and shit represented scientifically as the gut microbiome. Autointoxication clearly recapitulated conventional discourses on the perils of base matter, fears of contact and contamination, apprehensions of internal colon-ization, the dire consequences of the dialectic of civilized and primitive, the pure and the dangerous, white and brown. Whereas the human gut microbiome is supposed to signal ecological relations, multispecies comity, carnivalesque colonic conviviality, promiscuous and beneficial boundary crossings, the self's tolerance or even incorporation of others. One gets an impression of a contrast between an abhorrence of base matter lurking within and a new respect

and reverence for our gut buddies. Then again, I could empha-
size the striking similarities of these two conceptual frames.
Although sometimes deemed cooperatively facilitative and
stabilizing, the gut microbiome still is relentlessly objectified
and imagined as separable from the human body, just like
poisonous feces once were. Despite ecological pretensions,
the microbiome in practice still is exposed as a euphemism
for the old shit, only now with a slightly divergent refiguring,
an ontological adjustment. In the early nineteenth century,
medicos sought to remove harmful intoxicating matter from
the body's innards, either by introducing microbial competitors, by colonic irrigation, or by resection of the large bowel.
These days, we also seek to allay anxieties about shit – to
distance ourselves from it – through massive investment in its
scientific inscription, though molecularization and agreeable
reanimation as the companionable microbiome. What cannot
be excreted or voided must be sublimated or rarified. Thus,
we witness analogous processes of fetishization and securitization, which supposedly allow safe mobilization of feces, not as
abject disturbing matter but rather, in the case of the human
gut microbiome, as a medicinal object, with at least promissory
value. Shit, when written up as an assemblage of microbial
genomes, evidently has little odor.

These days, human waste is more likely imagined as con-
figuration, not contamination, but all the same, the customary
tropes that position modernity against nature, civilization
against primitivism, keep returning to inform it. We continue to resist reasoning our excrement ecologically. Shit is
thus repeatedly subjected to re-colon-ization – not conjured
as counter-biopolitics. To be sure, "primitive" intestinal biota
may now appear healthier and more beneficially diverse, but
this nativized shit can be safely analyzed only in the sterile
and controlled conditions of a scientific laboratory, where it
is manipulated into a secure dataset. Few white people, aside
from the odd Texan adventurer, are happy to embrace raw

hunter-gatherer feces. They can only contemplate shit amiably when it is sanitized and sublimated, rendered inter-operable and transactional – even encapsulated – whitened through the ritual frames of a hygienic laboratory. There, all that is solid melts into microbiota. Only when feces are rendered calculable and standardized do they move on, from a state of exception, from bare life or base matter, to attain certified ratings of bioavailability[72] – only then can they be accepted, sometimes, as tolerable. Short of any such scientific accreditation, needy stool recipients like Tracy are compelled to turn to that other icon of hygienic modernity, the athletic white-male super-donor to ensure safe re-colon-ization. So much for re-wilding the deficient modern colon. So much for becoming ecological holobionts. So much, then, for living agreeably with shit, let alone de-colon-izing it.

Conclusion

A Topsy-Turvy Creature

In *Crowds and Power* (1960), the European writer Elias Canetti, born in Bulgaria and brought up in post-Kakania Austria between the wars, expatiated in his usual eloquent but paranoid fashion on digestion and defecation:

> The relation of each and every man to his own excrement belongs to the sphere of power. Nothing has been so much part of one as that which turns into excrement. The constant pressure, which, during the whole of its long progress through the body, is applied to the prey which has become food; its dissolution and intimate union with the creature digesting it; the complete and final annihilation, first of all functions and then of everything which once constituted its individuality; its assimilation to something already existing, that is, the body of the eater – all this may very well be seen as the central, if most hidden, process of power.[1]

As he considered the excremental workings of biopower, Canetti observed that Europeans take great care to secrete their waste, to disavow their bowels:

It is remarkable how we isolate ourselves with it; in special rooms, set aside for the purpose, we get rid of it; our most private moment is when we withdraw there; we are alone only with our own excrement. It is clear we are ashamed of it. It is the age-old seal of that power-process of digestion, which is enacted in darkness and which, without this, would remain hidden forever.[2]

It has been my purpose here to bring to light these peristaltic intimacies of power, to lay out for inspection the excreted evidence of how, for centuries, shit has stimulated and challenged modern reason, bioscience, and aesthetics. As we have seen, anxieties over base matter, fears of contamination and degradation, worries about pollution and disorder, have tested and renovated our sense of ourselves, added definition to modern bodies, reshaped population sciences like epidemiology, prompted philosophy, psychoanalysis, and social theory, and brought forth literature and art. Science and art are, to a degree, symptomatic of our captivation with shit and our aversion to it. As W.H. Auden, following Sigmund Freud, put it:

All the arts derive from
This ur-act of making,
Private to the artist.[3]

It is almost as though speculation on antithetical shit is necessary to keep us thinking we are modern and civilized. Defecation thus turns us into "modern" humans – or at least, we sense the act of voidance and expulsion might bring us closer, asymptotically, toward modernity.

In a roundabout way, as though stepping delicately through the streets of Jonathan Swift's Edinburgh, I have tried to detect various modes of removing or warding off shit, that archetypical matter out of place, perceived as a fundamental threat to order and security. I have concentrated on biomedical reasoning about shit, on efforts to render it calculable and safe, even informative and beneficial, through wastewater epidemiology and molecular translations of the gut microbiome, the case studies that bookend my narrative. But I have also conducted excursions that reveal shit's salience in psychoanalysis, social theory, cultural anthropology, literature, and art, if only to allow some comparison with its significance in these other domains of modern human thought. Predictably, perhaps, my exploration of modern excrementalities has dredged up the special (though not singular) proclivity of white men in or from the northeastern United States and Western Europe for making such fecal distinctions, for fixating on ontologizing and securitizing shit, for venturing out to colon-ize the world. In periods of high anxiety, times when fears of the porosity of boundaries are particularly intense such as the early twentieth century and the twenty-first century, these white men have ferociously accentuated and extended the biopolitical signature of shit – though now they are no longer alone in so doing.

My general argument should be evident. Time and again, we take recourse to structural dichotomies, such as purity and danger, modern and primitive, culture and nature, to organize ourselves against disorderly shit, to distance ourselves from it and to make it invisible or innocuous. We seem incapable of escaping this excremental dialectic, even when we seek to romance the turd, to celebrate dirt and fecal liveliness, even when we offer somewhat futile gestures toward alternative ecological reasoning. Within the insistent binary frame, we keep resorting to a range of apotropaic genres, sublimations or displacements that promise to avert or interdict excremental damage, to turn the abject into a mere object with which we can live securely. But such inscription devices and data methods prove never to be enough; they break down, fall short, allowing the return of the repressed. The stool might be made a fetish, but as with all fetishes, we know it is never quite what we would have it appear to be; rather, it is always ambiguous, disingenuous, and assailable. And so, as Bruno Latour told us, we find out again and again that we are not really modern – that perhaps we never *can* be modern.

At the same time, this is also a story of the spectacularizing of human waste – or more precisely, the spectacularizing of the anal speculum – making feces widely visible without being directly grasped. It is a litany of efforts, often vain, to render shit nothing more than abstraction and alienation – especially in the clean spare space of the modern laboratory. Accordingly, stool in commodity form is mobilized in order to colon-ize all social life – though this conceit may never quite be achieved. As Situationist Guy Debord, following Karl Marx, claimed, "the fetishistic appearance of pure objectivity in spectacular relations conceals their true character as relations between people and between classes [or classifications]." Our modern shit spectacle works by "incorporating into itself all the *fluid* aspects of human activity so as to possess them in congealed form."[4] Or as Jean Baudrillard expressed more candidly: in a consumer society everything "is finally digested and reduced to the same homogeneous fecal matter . . . a controlled, lubricated, and consumed excretion (*fécalité*) is hence transferred into things, everywhere diffused in the indistinguishability of things and social relations."[5] We alleged moderns, those deemed developed or civilized, cannot, so it seems, avoid getting caught up in this spectacle of waste, with few options for what the Situationists called *détournement*, reframing objectified shit so as to subvert its authority, perhaps to ecologize it.

A related theme of this book is the mapping of those persistent binaries of purity and danger, or hygiene and filth, onto colonial relations – or rather, the spectacular expression of such binaries through colonial bodies. Thus, we see white medical experts and public health officers displace concerns about excrementality onto other races, particularly brown and black, transforming them into fearsomely promiscuous defecators, while attempting, unsustainably, to expunge or nullify their own European lower bodily stratum. This process of colon-izing other races, the poor, and the marginalized represents more explicitly than any other disciplinary program the mate-riality, the physicality, of the colonial project. Such colon-izing of others is surely among the more spectacular performances of colonial governmentality. But ultimately no one is immune from the creeping fecal signifier. As we have seen, one does not have to be a colonial subject to be transmuted into an excremental object. Indeed, through the twentieth century, white folk, and newly "developed" people, have repeatedly been drawn back, often reluctantly, to colon-izing themselves, imagining their intestinal contents as a fascinating yet scary dark continent, a place within requiring constant surveillance and documentation, a site demanding relentless datafication and control. This implacable return of excrement therefore requires us to confront the decolonial limit experience – the rule that we cannot ever conceive of breaking – an impasse so intense and a voidance so paradoxical that we feel we must resist, in face of shit, challenging the entrenched colon-ial order of things. We refuse to think outside the latrine. The rectum, then, is the site where decolonization becomes inexpressible if not unimaginable. It seems we erstwhile moderns dare not ever be truly de-colon-ized.

When it comes to thinking otherwise about shit, to imag-ining life with excrement beyond the boundaries, outside conventional typologies, as complexly ecological, I admit I too am at a loss, stranded like any would-be modern in the deco-lonial limit experience. But I hope this story at least has shed some light on the significance of the privy, with its hidden inti-macies of biopower, and on the multiple meanings of human defecation in the modern era, allowing us thus to see more clearly our self-constitutions and our insecurities, our libidi-nal investments in knowledge making and in evaluation of its products. It seems Swift may have been there first:

Pray what is Man, but a Topsy-turvy Creature, his Animal Faculties perpetually a Cock-Horse and Rational; his Head where his Heals should be; groveling on the Earth, and yet with all his faults, he sets up to be a universal Reformer and Corrector of Abuses, a Remover of grievances, rakes into every Slut's Corner of Nature, bringing hidden Corruptions to the Light, and raises a mighty Dust where there was none before, sharing deeply all the while in the very same Pollutions he pretends to sweep away.[6]

Maybe Swift was right after all.

Acknowledgments

For almost thirty years, I have been writing intermittently about shit – some may say "writing shit" – all the while fearing that one day I will acquire as a result an embarrassing "courtesy stigma."[1] Perhaps that moment finally has arrived. In any case, I am especially grateful to my family and to Hans Pols for putting up with such preoccupations. Sometimes, my family went further than mere tolerance: until his death in 2017, my father, Hugh Anderson, would ply me with scatological materials from eighteenth- and nineteenth-century British popular culture.

During three decades of resolute shit stirring, I have acquired many intellectual debts, too many to recall completely. I expect the influence of my teachers will be obvious: Charles Rosenberg, my Ph.D. advisor long ago; Bruno Latour (at Melbourne in 1988); and Homi Bhabha (at Penn in 1994). When I taught at Harvard in the 1990s, Mary Steedly, Vince Rafael, Begoña Aretxaga, Gabrielle Hecht, Michelle Murphy, and Anna Tsing urged me to write on excremental colonialism, which led to an article with that title published in *Critical Inquiry*. My colleagues in the Department of the History of Science were bemused but showed exemplary open-mindedness.

Later at UCSF and Berkeley, Adele Clarke, Philippe Bourgois, Lawrence Cohen, Tom Laqueur, Sharon Kaufman, Gabriela Soto Laveaga, James Vernon, Paul Rabinow, Linda Mitteness, and Judith Barker shaped my thinking on excretory matters, in the nicest possible way. I still recall the moment in the Lower Haight when Nick King enthusiastically gifted me the phrase, "crap on the map."

At the University of Sydney, I've benefited greatly from discussions with Sophie Chao, James Dunk, Sheila Fitzpatrick, Sebastián Gil-Riaño, Paul Griffiths, Dirk Moses, Maureen O'Malley, Tamson Pietsch, Hans Pols, Elspeth Probyn, Kane Race, and Sonja van Wichelen. Sam Widin and Genevieve Dally-Watkins provided research assistance. From Lisa Adkins, Annemarie Jagose, Steve Simpson, Rodney Smith, and Lee Wallace, I received crucial institutional support and protection while writing. Indeed, I couldn't imagine a better environment for composing a book on shit, which I mean as high praise, of course. From UNSW across the city, Kari Lancaster offered helpful advice on wastewater epidemiology. Cathy Waldby, Ken George, and Kirin Narayan gave me encouragement from Canberra, as did Byron Good from Jogjakarta. From Charleston, South Carolina, Jacob Steere-Williams kept pushing me to take an even sharper excremental turn. In Rozelle, Simon Smith directed me to salient examples of shit in contemporary popular culture. Dhiraj Kumar Mohan Nainani advised me on practices of wastewater epidemiology in Singapore. And Jeremy Greene and Anne Kveim Lie showed remarkable forbearance when my excretions distracted me from our other projects.

A group of friends whom I've known since earlier days in Melbourne, before this shit was even a twinkle in my eye, assisted in framing my research and inspiring its completion. I'm grateful to Jill Carrick, Dipesh Chakrabarty, Nick Haslam, Matthew Klugman, Emma Kowal, Felicity Scott, Mark Veitch, and Fiona Wilson. Mark and Fiona stimulated my thinking on the relations of shit to COVID-19, though they don't always agree with my inferences.

Above all, I express heartfelt thanks to Philippa Barr, Margaret Jolly, Hans Pols, and Sonja van Wichelen, who read every word of the manuscript and commented extensively. Maureen O'Malley delved deep into the bowels of Chapter 5. I just wish I could have responded more adequately to their brilliant suggestions. Alas, the multispecies shit chapter, focused on wombat scat, that Hans wanted will remain unwritten.

Unless otherwise attributed, I am responsible for French translations; and Hans Pols did the Dutch translations.

While not something commonly spoken about in public, shit was the subject of some talks I gave, which helped me get it together. I limited these to one talk each decade: a seminar in the Department of Anthropology at UC Santa Cruz in 1994; a paper for Sarah Hodges's "Biotrash" conference at Warwick University's palazzo in Venice (which I insisted was really *my* palazzo) around 2009; and for Annabel Kim in her "Close Encounters of the Fecal Kind" workshop at the Harvard Radcliffe Institute in 2018. Both Sarah and Annabel have been generous interlocutors. Laura Meek's invitation to be a discussant for her panel "Multivalent Shit: New Directions in the Anthropology of Excrement, Defecation, and Manure," at the 2021 annual meeting of the American Anthropological Association, stimulated further interest in the subject, though unfortunately I could not attend, not even online. My articles on the topic have been published in *Critical Inquiry* (1995) and *Postcolonial Studies* (2010) and *Somatosphere* (2022): I am grateful to the reviewers and editors of those publications, especially John Cash and Eugene Raikhel. At Polity Press, Julia Davies spotted the potential for a book, then carefully and dexterously guided me through the writing process, making it all happen.

I acknowledge, too, that I wrote this book on the land of the Wangal of the Eora nation, in what is now called Sydney, and on an island in Dyarubbin (Hawkesbury River) on Guringai land.

Notes

Introduction

1   Emma Garnett, Angeliki Balayannis, Steve Hinchliffe, Thom Davies et al., "The work of waste during COVID-19: Logics of public, environmental, and occupational health," *Critical Public Health* (2022): https://doi.org/10.1080/09581596.2022.2 048632.

2   Judith Butler, *What World Is This? A Pandemic Phenomenology* (New York: Columbia University Press, 2022), pp. 33, 45.

3   David M. Berendes, Patricia J. Yang, Amanda Lai, David Hu, and Joe Brown, "Estimation of global recoverable human and animal faecal biomass," *Nature Sustainability* 1 (2018): 679–85.

4   James L.A. Webb, Jr., *The Guts of the Matter: A Global History of Human Waste and Infectious Intestinal Disease* (Cambridge: Cambridge University Press, 2018).

5   Steven Shapin and Simon Schaffer, *Leviathan and the Air-Pump: Hobbes, Boyle, and the Experimental Life* (Princeton: Princeton University Press, 1985).

6   Dominique Laporte, *A History of Shit* [1978], trans. Rodolphe El-Khoury (Cambridge MA: MIT Press, 2000), p. 65.

7   Roy Porter, "The patient's view: Doing medical history from below," *Theory and Society* 14 (1985): 175–85.

8   Friedrich Nietzsche, *The Genealogy of Morals: An Attack* [1887], trans. Francis Golffing (Garden City NY: Doubleday and Co., 1956), p. 160.

9   Bruno Latour, *We Have Never Been Modern*, trans. Catherine Porter (Cambridge MA: Harvard University Press, 1993).

10 Geoffrey C. Bowker and Susan Leigh Star, *Sorting Things Out: Classification and its Consequences* (Cambridge MA: MIT Press, 2000). See also Ian Hacking, "Making up people" [1986], in *Historical Ontology* (Cambridge MA: Harvard University Press, 2004), pp. 99–114.

11 Susan Leigh Star and Martha Lampland, "Reckoning with standards," in *Standards and Their Stories: How Quantifying, Classifying, and Formalizing Practice Shape Everyday Life* (Ithaca: Cornell University Press, 2009), pp. 3–14, p. 14. See Brian Larkin, "The politics and poetics of infrastructure," *Annual Review of Anthropology* 42 (2013): 327–43; and Peter Redfield and Steven Robins, "An index of waste: humanitarian design, 'dignified living' and the politics of infrastructure in Cape Town," *Anthropology Southern Africa* 39, 2 (2016): 145–62.

12 Michel Foucault, "The confession of the flesh" [1977], in *Power/ Knowledge: Selected Interviews and Other Writings*, trans. Colin Gordon et al. (New York: Pantheon, 1980), pp. 194–228.

13 Giorgio Agamben, *"What is an Apparatus?" and Other Essays*, trans. David Kishik and Stefan Pedatella (Stanford: Stanford University Press, 2009), p. 14.

14 Foucault, "The confession of the flesh," p. 195, original emphasis.

15 On rubbish and waste generally, see Michael Thompson, *Rubbish Theory: The Creation and Destruction of Value* (Oxford: Oxford University Press, 1979); Gay Hawkins, *The Ethics of Waste: How We Relate to Rubbish* (Lanham MD: Rowman and Littlefield, 2005); Gay Hawkins and Stephen Muecke, eds., *Culture and Waste: The Creation and Destruction of Value* (Lanham MD: Rowman and Littlefield, 2003); Sarah Hodges, "Medical garbage and the making of neo-liberalism in India," *Economic and Political Weekly* 48, 48 (2013): 112–19; and Marco Armiero,

Wasteocene: Stories from the Global Dump (Cambridge: Cambridge University Press, 2021). On discard theory, see Kate O'Neill, *Waste* (Cambridge: Polity, 2019); Max Liboiron and Josh Lepawsky, *Discard Studies: Wasting, Systems, and Power* (Cambridge MA: MIT Press, 2022); Patrick O'Hare, *Rubbish Belongs to the Poor: Hygienic Enclosure and the Waste Commons* (London: Pluto Press, 2022).

16 Sarah A. Moore, "Garbage matters: Concepts in new geographies of waste," *Progress in Human Geography* 36, 6 (2012): 780–99; and Max Liboiron, *Pollution is Colonialism* (Durham NC: Duke University Press, 2021).

17 According to Christian Enzensberger, "even written matter can also be conceived of as excretion" (*Smut: An Anatomy of Dirt*, trans. Sandra Morris [London: Calder and Boyars, 1972], p. 34).

Chapter 1 The Sewage Panopticon

1 Don DeLillo, *Underworld* (New York: Scribner, 1997), p. 103.

2 DeLillo, *Underworld*, pp. 344, 326.

3 Janelle Thompson, Yarlagadda V. Nancharaiah, Xioaqiong Gu, Wei Lin Lee et al., "Making waves: Wastewater surveillance of SARS-CoV-2 for population-based health management," *Water Research* 184 (2020): https://www.sciencedirect.com/science/article/pii/S0043135420307181.

4 On the sewage infrastructure crisis, see Ian A. Wright, "If the tide is high, our sewerage systems won't hold on," *The Conversation* (2013): https://theconversation.com/if-the-tide-is-high-our-sewerage-systems-wont-hold-on–14467; Catherine Coleman Flowers, *Waste: One Woman's Fight Against America's Dirty Secret* (New York: New Press, 2020); Elisa Gambino, *Wasteland* [TV miniseries], Paramount Plus (2022); Oliver Bullough, "Sewage sleuths: The men who revealed the slow, dirty death of Welsh and English rivers," *Guardian* (2022): https://www.theguardian.com/environment/2022/aug/04/sewage-sleuths-river-pollution-slow-dirty-death-of-welsh-and-english-rivers; and Christopher Flavelle, "A stinky stew on Cape Cod: Human

waste and warm water," *New York Times* (2023): https://www
.nytimes.com/2023/01/01/climate/cape-cod-algae-septic.html.
For effects of global heating, see Jonathan A. Patz, Stephen J.
Vavrus, Christopher K. Uejio, and Sandra L. McLellan, "Climate
change and waterborne disease risk in the Great Lakes region of
the U.S.," *American Journal of Preventive Medicine* 35, 5 (2008):
451–58, p. 453.

5 Charles Bramesco, "'This is everybody's problem': Inside
America's growing sewage crisis," *Guardian* (2022): https://www
.theguardian.com/tv-and-radio/2022/feb/28/wasteland-amer
icas-growing-sewage-crisis-docuseries.

6 Edward Segal, "Biden administration seeks to bolster defenses
against cyberattacks on water systems," *Forbes* (2022): https://
www.forbes.com/sites/edwardsegal/2022/02/13/bidden-admini
stration-seeks-to-bolster-defenses-against-cyberattacks-on-
water-systems/?sh=29240f3c1ff9.

7 Marine Legrand and German Meulemans, "Bathing in black
water? The microbiopolitics of the River Seine's ecological
reclamation," in *With Microbes*, ed. Charlotte Brives, Matthäus
Rest and Salla Sariola (Manchester: Mattering Press, 2021),
pp. 143–64.

8 William James, *The Varieties of Religious Experience: A Study in
Human Nature* (New York: Longmans, Green, 1902).

9 Aldous Huxley, *Do What You Will* (London: Chatto and Windus,
1931), p. 94.

10 Willemijn Lodder and Anna Maria de Roda Husman, "SARS-
CoV-2 in wastewater: Potential health risk, but also data source,"
The Lancet: Gastroenterology and Hepatology 5, 6 (2020): 533–4;
Gertjan Medema, Leo Heinen, Goffe Elsinga, Ronald Italiaander
et al., "Presence of SARS-Coronavirus–2 RNA in sewage and cor-
relation with reported COVID-19 prevalence in the early stage
of the epidemic in the Netherlands," *Environmental Science and
Technology Letters* 7, 7 (2020): 511–16; F.Q. Wu, J.B. Zhang,
A. Xiao, X.Q. Gu et al., "SARS-CoV-2 titers in wastewater are
higher than expected from clinically confirmed cases," *mSystems*

5, 4 (2020): https://doi.org/10.1128/mSystems.00614–20; Warish Ahmed, Nicola Angel, Janette Edson, Kyle Bibby et al., "First confirmed detection of SARS-CoV-2 in untreated wastewater in Australia: A proof of concept for the wastewater surveillance of COVID-19 in the community," *Science of the Total Environment* 728 (2020): https://doi.org/10.1016/j.scitotenv.2020.138764; Manish Kumar, Armin Kumar Patel, Anil V. Shah, Janvi Raval et al., "First proof of the capability of wastewater surveillance for COVID-19 in India through detection of genetic material of SARS-CoV-2," *Science of the Total Environment* 746 (2020): https://doi.org/10.1016/j.scitotenv.2020.141326; and Renee Street, Shirley Malema, Nomfundo Mahlangeni, and Angela Mathee, "Wastewater surveillance for Covid-19: An African perspective," *Science of the Total Environment* 743 (2020): https://doi.org/10.1016/j.scitotenv.2020.140719.

11 Smriti Mallapaty, "How sewage could reveal true scale of coronavirus outbreak," *Nature* 580 (2020): 176–7, p. 176.

12 Manzoor Qadir, "Needed in the Global South: Wastewater collection for COVID-19 detection," *Our World* (2021): https://ourworld.unu.edu/en/needed-in-the-global-south-wastewater-collection-for-covid–19-detection.

13 Freda Kreier, "The myriad ways sewage surveillance is helping fight COVID around the world," *Nature* (2021): https://www.nature.com/articles/d41586–021–01234–1.

14 Pubali Mandal, Ashok K. Gupta, and Brajesh K. Dubey, "A review on presence, survival, disinfection/removal methods of coronavirus in wastewater and progress of wastewater-based epidemiology," *Journal of Environmental Chemical Engineering* 8, 5 (2020): https://doi.org/10.1016/j.jece.2020.104317.

15 Sahana Ghosh, "Can human poop help track the spread of coronavirus in Indian localities?" *Quartz* (2020): https://qz.com/india/1837014/can-human-poop-help-track-the-spread-of-coronavirus-in-indian-localities.

16 David Graeber, *Bullshit Jobs: A Theory* (New York: Simon and Schuster, 2018).

17 On scriptural economies, see Michel De Certeau, *The Practice of Everyday Life*, trans. S.F. Rendall (Berkeley: University of California Press, 1984).

18 Bruno Latour, "Visualization and cognition: Thinking with eyes and hands," in *Knowledge and Society: Studies in the Sociology of Culture Past and Present* 6, ed. Henrika Kuklick and Elizabeth Long (Greenwich CT: JAI Press, 1986), pp. 1–40, p. 17.

19 Latour, "Visualization and cognition," p. 32.

20 DeLillo, *Underworld*, p. 326.

21 Brady Dennis, "An early warning system for coronavirus infections could be found in your toilet," *Washington Post* (2020): https://www.washingtonpost.com/climate-environment/2020/05/01/coronavirus-sewage-wastewater; and Celia Henry Arnaud, "Weighing wastewater's worth as a COVID-19 monitoring tool," *Chemical and Engineering News* (2021): https://cen.acs.org/biological-chemistry/infectious-disease/Weighing-wastewaters-worth-COVID-19/99/i35.

22 Matt Stieb, "How wastewater became a COVID crystal ball," *Intelligencer* (2022): https://www.msn.com/en-us/news/technology/how-wastewater-became-a-covid-crystal-ball/ar-AASOnvO.

23 Emily Anthes, "In New York City sewage, a mysterious coronavirus signal," *New York Times* (2022): https://www.nytimes.com/2022/02/03/health/coronavirus-wastewater-new-york.html; and Davida S. Smyth, Monica Trujillo, Devon A. Gregory, Kristen Cheung et al., "Tracking cryptic SARS-CoV-2 lineages detected in NYC Wastewater," *Nature Communications* 13 (2022): https://doi.org/10.1038/s41467-022-28246-3.

24 Apoorva Mandavilli, "The C.D.C. isn't publishing large portions of the Covid data it collects," *New York Times* (2022): https://www.nytimes.com/2022/02/20/health/covid-cdc-data.html. See Centers for Disease Control and Prevention, "National wastewater surveillance system (NWSS)," (2020): https://www.cdc.gov/healthywater/surveillance/wastewater-surveillance/wastewater-surveillance.html.

25 Maarten Keulemans, "Het riool wordt leidraad bij het volgen van het coronavirus," *De Volkeskrant* (2022): https://www.volkskrant.nl/wetenschap/het-riool-wordt-leidraad-bij-het-volgen-van-het-coronavirus~ba20f66e.

26 Warish Ahmed, Paul M. Bertsch, Nicola Angel, Kyle Bibby et al., "Detection of SARS-CoV-2 RNA in commercial passenger aircraft and cruise ship wastewater: A surveillance tool for assessing the presence of COVID-19 infected travellers," *Journal of Travel Medicine* (2020): https://doi.org/10.1093/jtm/taaa116; and David S. Thaler and Thomas P. Sakmar, "Archiving time series sewage samples as biological records of built environments," *BMC Infectious Diseases* 21 (2021): 601.

27 Natalie Sims and Barbara Kasprzyk-Hordern, "Future perspectives of wastewater-based epidemiology: Monitoring infectious disease spread and resistance to the community level," *Environment International* 139 (2020): https://doi.org/10.1016/j.envint.2020.105689; and D.A. Larsen and K.R. Wigginton, "Tracking COVID-19 with wastewater," *Nature Biotechnology* 38 (2020): 1151–3.

28 Keulemans, "Het riool wordt leidraad bij het volgen van het coronavirus."

29 T. Jessie Ge, Carmel T. Chan, Brian J. Lee, Joseph C. Liao, and Seung-min Park, "Smart toilets for monitoring COVID-19 surges: Passive diagnostics and public health," *NPJ Digital Medicine* 5, 39 (2022): https://doi.org/10.1038/s41746–022–00582–0.

30 Jacob Steere-Williams, "Endemic fatalism and why it will not resolve COVID-19," *Public Health* 206 (2020): 29–30.

31 For the early medical history of excrement, see Shigehisa Kuriyama, "The forgotten fear of excrement," *Journal of Medieval and Early Modern Studies* 38, 3 (2008): 413–42.

32 Charles E. Rosenberg, *The Cholera Years: The United States in 1832, 1849, and 1866* (Chicago: University of Chicago Press, 2008).

33 On sewerage, see Christopher Hamlin, "Edwin Chadwick and the engineers, 1842–1854: Systems and antisystems in the

pipe-and-brick sewers," *Technology and Culture* 33, 4 (1992): 680–709; Joseph W. Childers, "Foreign matter: Imperial filth," in *Filth: Dirt, Disgust and Modern Life*, ed. William A. Cohen and Ryan Johnson (Minneapolis: University of Minnesota Press, 2005), pp. 201–21; Michelle Allen, *Cleansing the City: Sanitary Geographies in Victorian London* (Athens OH: Ohio University Press, 2008); and Jacob Steere-Williams, *The Filth Disease: Typhoid Fever and the Practices of Epidemiology in Victorian England* (Rochester NY: University of Rochester Press, 2020). On flush toilets, see Virginia Sarah Smith, *Clean: A History of Personal Hygiene and Purity* (Oxford: Oxford University Press, 2008); and Slavoj Žižek, "Knee deep," *London Review of Books* 26, 17 (2004): 12–13.

34 Edward Playter, "Our inland lakes and rivers, the disposal of sewage, and the spread of infectious diseases," *Public Health Papers and Reports* 12 (1886): 123–32, p. 123.

35 Joel A. Tarr and Gabriel Dupuy, *Technology and the Rise of the Networked City in Europe and America* (Philadelphia: Temple University Press, 1988); Donald Reid, *Paris Sewers and Sewermen: Realities and Representations* (Cambridge MA: Harvard University Press, 1993); David L. Pike, "Sewage treatments: Vertical space and waste in nineteenth-century Paris and London," in *Filth*, ed. Cohen and Johnson, pp. 51–77; and Martin V. Melosi, *The Sanitary City: Environmental Services in Urban America from Colonial Times Until the Present* (Pittsburgh: University of Pittsburgh Press, 2008).

36 Erwin H. Ackerknecht, *Rudolf Virchow: Doctor, Statesman, Anthropologist* (Madison WI: University of Wisconsin Press, 1953).

37 Milan Kundera, *The Unbearable Lightness of Being*, trans. Michael Henry Heim (London: Faber and Faber, 1984), p. 151.

38 Steven J. Burian, Stephan J. Nix, Robert E. Pitt, and S. Rocky Durrans, "Urban wastewater management in the United States: Past, present, and future," *Journal of Urban Technology* 7, 3 (2000): 33–62; and Jamie Bendrickson, *The Cultures of Flushing:*

A Social and Legal History of Sewage (Vancouver: University of British Columbia Press, 2007).

39 C-E.A. Winslow and D.M. Belcher, "Changes in the bacterial flora of sewage during storage," *Journal of Infectious Diseases* 1 (1904): 170–92.

40 Stephen D. Gage and George O. Adams, "The collection and preservation of samples of sewage for analysis," *Journal of Infectious Diseases* 3, Supplement 2 (1906): S139–S148.

41 Warwick Anderson, "Think like a virus," *Public Books* (2021): https://www.publicbooks.org/think-like-a-virus.

42 Lawrence Goodridge, "Sewage surveillance: How scientists track and identify diseases like COVID-19 before they spread," *The Conversation* (2020): https://theconversation.com/sewage-surveillance-how-scientists-track-and-identify-diseases-like-covid–19-before-they-spread–148307.

43 J.R. Paul, J.D. Trask, and S.I. Gard, "Poliomyelitic virus in urban sewage," *Journal of Experimental Medicine* 71, 6 (1940): 765–77.

44 John T. Riordan,. "Isolation of enteroviruses from sewage before and after vaccine administration," *Yale Journal of Biology and Medicine* 34, 5 (1962): 512–21.

45 Badri Fattal and Moshe Nishmi, "Enterovirus types in Israel sewage," *Water Research* 11, 4 (1977): 393–6.

46 World Health Organization, *Guidelines for Environmental Surveillance of Polio Circulation* (Geneva: WHO, 2003).

47 X.W. Wang, J. Li, T. Guo, B. Zhen et al., "Concentration and detection of SARS coronavirus in sewage from Xiao Tang Shan Hospital and the 309th Hospital of the Chinese People's Liberation Army," *Water Science and Technology* 52 (2005): 213–21.

48 Christian C. Daughton, "Illicit drugs in municipal sewage: Proposed new nonintrusive tool to heighten public awareness of societal use of illicit-abused drugs and their potential for ecological consequences," in *Pharmaceuticals and Care Products in the Environment: Scientific and Regulatory Issues*, ed. Christian

C. Daughton and Tammy L. Jones-Lepp (Washington DC: American Chemical Society, 2001), pp. 348–64: https://doi.org /10.1021/bk–2001–0791.ch020.

49 Ettore Zuccato, Chiara Chiabrando, Sara Castiglioni, Davide Calimari et al., "Cocaine in surface waters: A new evidence-based tool to monitor community drug abuse," *Environmental Health* 4 (August 2005): https://doi.org/10.1186/1476–069X–4–14.

50 Kari Lancaster, Alison Ritter, kylie valentine, and Tim Rhodes, "'A more accurate understanding of drug use': A critical analysis of wastewater analysis technology for drug policy," *International Journal of Drug Policy* 63 (2019): 47–55, p. 47.

51 Kari Lancaster and Tim Rhodes, "Wastewater monitoring for SARS-CoV-2: Lessons from illicit drug policy," *The Lancet: Gastroenterology and Hepatology* 5, 7 (2020): 641–2, p. 641.

52 Christian C. Daughton, "Wastewater-based epidemiology: A 20-year journey may pay off for Covid-19," *Stat* (2021): https:// www.statnews.com/2021/01/07/wastewater-based-epidemi ology–20-year-journey-pay-off-for-covid-19.

53 Eric Hagerman, "Your sewer on drugs," *Popular Science* (2008): https://www.popsci.com/scitech/article/2008–02/your-sewer -drugs; and Carrie Arnold, "Pipe dreams: Tapping into the health information in our sewers," *Environmental Health Perspectives* 124, 5 (2016): A86–91.

54 Michel Foucault, *Security, Territory, Population: Lectures at the Collège de France, 1977–1978*, trans. Graham Burchell (Basingstoke: Palgrave Macmillan, 2009).

55 T.G. Aw and K.Y.-H. Gin, "Environmental surveillance and molecular characterization of human enteric viruses in tropical urban wastewaters," *Journal of Applied Microbiology* 109, 2 (2010): 716–30.

56 Thai-Hoang Le, Charmaine Ng, Hongjie Chen, Xin Zhu Yi et al., "Occurrences and characterization of antibiotic-resistant bacteria and genetic determinants of hospital wastewater in a tropical country," *Antimicrobial Agents and Chemotherapy* 60, 12 (2016): https://doi.org/10.1128/AAC.01556–16.

57 Franciscus Chandra, Wei Lin Lee, Federica Armas, Mats Leifels et al., "Persistence of dengue (serotypes 2 and 3), Zika, yellow fever, and murine hepatitis virus RNA in untreated wastewater," *Environmental Science and Technology Letters* 8, 9 (2021): 785–9.

58 National Environment Agency, "NEA leads scientific team in wastewater surveillance trials for assessment of COVID-19 transmission" (2020): https://www.nea.gov.sg/media/news/news/index/nea-leads-scientific-team-in-wastewater-surveillance-trials-for-assessment-of-covid–19-transmission; and Janelle Thompson, Yarlagadda V. Nancharaiah, Xioaqiong Gu, Wei Lin Lee et al., "Making waves: Wastewater surveillance of SARS-CoV-2 for population-based health management," *Water Research* 184 (2020): https://www.sciencedirect.com/science/article/pii/S0043135420307181.

59 Matthew Mohan, "From manhole to sampling bottle: How wastewater helps indicate presence of COVID-19 in foreign worker dormitories," *Channel News Asia* (2020): https://www.channelnewsasia.com/news/singapore/foreignworker-dormitories-sampling-testing-covid-19-wastewater–12953408.

60 Stephanie Armour and Brianna Abbott, "Biden unveils new Covid-19 strategy for next phase of response," *Wall Street Journal* (2022): https://www.wsj.com/articles/biden-unveils-new-covid-19-strategy-for-next-phase-of-response–11646237143.

61 White House, *National COVID-19 Preparedness Plan* (2022): https://www.whitehouse.gov/covidplan, p. 12.

62 Dominique Laporte, *A History of Shit* [1978], trans. Rodolphe El-Khoury (Cambridge MA: MIT Press, 2000), p. 66. On "technologies of distance" and "mechanical objectivity," see Ted Porter, *Trust in Numbers: The Pursuit of Objectivity in Science and Public Life* (Princeton: Princeton University Press, 1995).

63 Warwick Anderson, "Objectivity and its discontents," *Social Studies of Science* 43 (2013): 557–76.

64 Laporte, *A History of Shit*, p. 118.

65 Roland Barthes, *Sade, Fourier, Loyola* (Paris: Seuil, 1971), p. 140.

66 Michel Foucault, *"Society Must be Defended": Lectures at the Collège de France, 1975–1976*, trans. David Macey (New York: Picador, 2003).

67 Aleksandr Solzhenitsyn, *The Gulag Archipelago* [1973–76], trans. Thomas P. Whitney and Harry Willetts (New York: Harper and Row, 1985).

Chapter 2 The Waste That Therefore I Am?

1   Dan-el Padilla Peralta, "Waste," in *Liquid Antiquity*, ed. Brooke Holmes and Karen Marta (Geneva: DESTE Foundation for Contemporary Art, 2017), pp. 116–19, p. 116.

2   Padilla Peralta, "Waste," pp. 116, 118–19, 118.

3   Peter Sloterdijk, *Spheres, Volume 1: Bubbles. Microsphereology* [1998], trans. Wieland Hoban (Los Angeles: Semiotext(e), 2011), pp. 329, 331.

4   Bruno Latour, *We Have Never Been Modern*, trans. Catherine Porter (Cambridge MA: Harvard University Press, 1993), p. 47.

5   Dominique Laporte, *A History of Shit* [1978], trans. Rodolphe El-Khoury (Cambridge MA: MIT Press, 2000), p. 13.

6   William James, *The Varieties of Religious Experience: A Study in Human Nature* (New York: Longmans, Green, 1902), p. 113.

7   Alain Corbin, *The Foul and the Fragrant: Odor and the French Social Imagination*, trans. Miriam L. Kochan, Roy Porter, and Christopher Prendergast (Cambridge MA: Harvard University Press, 1986), p. 231.

8   Laporte, *A History of Shit*, pp. 44, 65.

9   Corbin, *The Foul and the Fragrant*, pp. 101, 158–9.

10 Corbin, *The Foul and the Fragrant*, pp. 143, 144. See also David S. Barnes, "Confronting sensory crisis in the great stinks of London and Paris," in *Filth: Dirt, Disgust and Modern Life*, ed. William A. Cohen and Ryan Johnson (Minneapolis: University of Minnesota Press, 2005), pp. 103–29.

11 Georges Vigarello, *Concepts of Cleanliness: Changing Attitudes in France Since the Middle Ages*, trans. Jean Birrell (Cambridge: Cambridge University Press, 1988).

12 Corbin, *The Foul and the Fragrant*, p. 227.

13 Florian Werner, "Dark matter: The history of shit," trans. Tereza Kuldora, *Journal of Extreme Anthropology* 1, 1 (2011): 63–80, pp. 65, 63.

14 Werner, "Dark matter," pp. 63, 65.

15 Norbert Elias, *The Civilizing Process: Sociogenetic and Psychogenetic Investigations* [1939], revised edition, trans. Edmund Jephcott (Oxford: Blackwell, 2000), p. 159.

16 W.H. Auden, "September 1, 1939," in *Another Time* (New York: Random House, 1940).

17 Sigmund Freud, "Character and anal eroticism" [1908], in *The Standard Edition of the Complete Psychological Works of Sigmund Freud*, trans. James Strachey, volume 9 (London: Hogarth Press, 1952), pp. 167–75.

18 Sigmund Freud, *Civilisation and Its Discontents* [1929], trans. Joan Rivière (London: Hogarth Press, 1973).

19 Robert Musil, *The Man Without Qualities* [1930], trans. Sophie Wilkins (New York: Picador, 1995).

20 Ernest Jones, "Anal-erotic character traits," *Journal of Abnormal Psychology* 13, 5 (1908): 261–84, p. 261.

21 Jones, "Anal-erotic character traits," pp. 269, 276, 278.

22 Abraham A. Brill, "Anal eroticism and character," *Journal of Abnormal Psychology* 7, 3 (1912): 196–203, pp. 203, 197.

23 Brill, "Anal eroticism and character," pp. 200, 197.

24 Karl Abraham, "Contributions to the theory of the anal character" [1921], in *Selected Papers on Psychoanalysis* (London: Routledge, 1927), pp. 370–92, pp. 377, 371, 375.

25 Abraham, "Contributions to the theory of the anal character," pp. 375, 377, 388.

26 Nick Haslam, *Psychology in the Bathroom* (Basingstoke: Palgrave Macmillan, 2012).

27 Erich S. Fromm, *Escape from Freedom* (New York: Farrer and Rinehart, 1941). See also Theodor Adorno, Else Frenkel-Brunswik, D.J. Levinson, and R.N. Stanford, *The Authoritarian Personality* (New York: Harper, 1950).

28 Erich S. Fromm, *To Have or To Be?* (New York: Harper and Row, 1976).

29 Erik Erikson, *Young Man Luther: A Study in Psychoanalysis and History* (New York: W.W. Norton, 1958), p. 199.

30 Herbert Marcuse, *Eros and Civilization: A Philosophical Inquiry into Freud* (Boston: Beacon Press, 1955).

31 Norman O. Brown, *Life Against Death: The Psychoanalytic Meaning of History* [1959], 2nd edition (Middletown CT: Wesleyan University Press, 1985), pp. 200, 180, 230.

32 Brown, *Life Against Death*, p. 230.

33 Theodore Roszak, *The Making of a Counter Culture: Reflections on the Technocratic Society and Its Youthful Opposition* (New York: Doubleday, 1969).

34 Mary Douglas, *Purity and Danger: An Analysis of Concepts of Pollution and Taboo* [1966] (London: Routledge, Kegan, Paul, 1976), pp. 120, 115.

35 Douglas, *Purity and Danger*, pp. 4, 134, 135.

36 Douglas, *Purity and Danger*, pp. 160, 165. On framing see Erving Goffman, *Frame Analysis: An Essay on the Organization of Experience* (Boston: Northeastern University Press, 1974).

37 Julia Kristeva, *Powers of Horror: An Essay in Abjection* [1980], trans. Leon S. Roudiez (New York: Columbia University Press, 1982), pp. 68, 71, 2, 4. See also Elizabeth Grosz, *Volatile Bodies: Toward a Corporeal Feminism* (Bloomington: Indiana University Press, 1994); Carol Wolkowitz, "Linguistic leakiness or really dirty? Dirt in social theory," in *Dirt: New Geographies of Cleanliness and Contamination*, ed. Ben Crampkin and Rosie Cox (I.B. Tauris, 2008), pp. 15–25; and Slavoj Žižek, "Abjection, disavowal and the masquerade of power," *Journal of the Centre of Freudian Analysis and Research* 26 (2015): 33–43.

38 Kristeva, *Powers of Horror*, pp. 4, 11.

39 Kristeva, *Powers of Horror*, pp. 11, 17, 26.

40 Michel Foucault, *The History of Sexuality, Vol. 1: An Introduction* [1971], trans. Robert Hurley (Harmondsworth: Penguin, 1981).

41 Sloterdijk, *Spheres, Volume 1: Bubbles*, pp. 410, 423.

42 Sloterdijk, *Spheres, Volume 1: Bubbles*, pp. 324, 327, 329, 325.
43 Sloterdijk, *Spheres, Volume 1: Bubbles*, pp. 330, 333.
44 Latour, *We Have Never Been Modern*, p. 11.
45 Latour, *We Have Never Been Modern*, pp. 115, 136.
46 Jacques Derrida, "White mythology: Metaphor in the text of philosophy" [1971], trans. Alan Bass, in *Margins of Philosophy* (Chicago: University of Chicago Press, 1982), pp. 209–71, p. 213.
47 Jacques Derrida, *The Beast and the Sovereign, Volume 1*, trans. Geoffrey Bennington (Chicago: University of Chicago Press, 2009), pp. 15, 18.
48 Jonathan Franzen, *The Corrections* [2001] (London: 4th Estate, 2022), pp. 282, 283.
49 Franzen, *The Corrections*, pp. 285, 286.
50 Sigmund Freud, "The uncanny" [1919], in *The Standard Edition of the Complete Psychological Works of Sigmund Freud*, trans. James Strachey, volume 17 (London: Hogarth Press, 1955), pp. 219–52, p. 249.

Chapter 3 The Colon-ized World

1 Toni Morrison, *Tar Baby* (New York: Vintage, 1981), pp. 183–4.
2 Morrison, *Tar Baby*, p. 184.
3 Morrison, *Tar Baby*, p. 184.
4 Jamaica Kincaid, *A Small Place* (London: Virago, 1988), p. 7.
5 Ann Laura Stoler, *Race and the Education of Desire: Foucault's History of Sexuality and the Colonial Order of Things* (Durham NC: Duke University Press, 2005); Anne McClintock, *Imperial Leather: Race, Gender, and Sexuality in the Colonial* Context (New York: Routledge, 1995); and Timothy Burke, *Lifebuoy Men, Lux Women: Commodification, Consumption, and Cleanliness in Modern Zimbabwe* (Durham NC: Duke University Press, 1996).
6 Warwick Anderson, "Excremental colonialism: Public health and the poetics of pollution," *Critical Inquiry* 21 (1995): 640–69.
7 Morrison, *Tar Baby*, p. 184.
8 V.S. Naipaul, *An Area of Darkness: An Experience of India* [1964] (New York: Picador, 1995), pp. 68, 70–1. For a twenty-first-

century account presuming a special affinity of Hindus with feces, see Gardiner Harrison, "Poor sanitation in India may afflict well-fed children with malnutrition," *New York Times* (2014): https://www.nytimes.com/2014/07/15/world/asia/poor-sanitation-in-india-may-afflict-well-fed-children-with-malnutrition.html – and critiques in Renu Desai, Colin McFarlane, and Stephen Graham, "The politics of open defecation: Informality, body, and infrastructure in Mumbai," *Antipode* 47 (2015): 98–120, and Assa Doron and Ira Raja, "The cultural politics of shit: Class, gender and public spaces in India," *Postcolonial Studies* 18, 2 (2015): 189–207. More attention should also be given to contemporary wasting of other "populations" (Michelle Murphy, "Against population, towards alterlife," in *Making Kin not Population*, ed. Adele E. Clarke and Donna Haraway [Chicago: Prickly Paradigm Press, 2018], pp. 101–24), especially of refugees (Kenan Malik, "Treating refugees like 'waste people' is abhorrent, wherever they end up," *Guardian* (2022): https://www.theguardian.com/commentisfree/2022/aug/21/treating-refugees-like-waste-people-is-abhorrent-wherever-they-end-up).

9 Dipesh Chakrabarty, "Of garbage, modernity and the citizen's gaze," *Economic and Political Weekly* 27, 10/11 (1992): 541–7, p. 541.

10 See also David Inglis, "Dirt and denigration: The faecal imagery and rhetorics of abuse," *Postcolonial Studies* 5, 2 (2022): 207–21; Alison Moore, "Colonial visions of 'Third World' toilets: A nineteenth-century discourse that haunts contemporary tourism," in *Ladies and Gents: Public Toilets and Gender*, ed. Olga Gershenson and Barbara Penner (Philadelphia: Temple University Press, 2009), pp. 105–25; and Max Liboiron, *Pollution is Colonialism* (Durham NC: Duke University Press, 2021).

11 Warwick Anderson, *Colonial Pathologies: American Tropical Medicine, Race, and Hygiene in the Philippines* (Durham NC: Duke University Press, 2006); and Vicente Rafael, *White Love and Other Events in Filipino History* (Durham NC: Duke University Press, 2000).

12 Bureau of Health, *The Disposal of Human Wastes in the Provinces*, Health Bulletin No. 13 (Manila: Bureau of Printing, 1912), pp. 4–5; DeWitt Wallace, "A few remarks concerning the health conditions of Americans in the Philippines," *Yale Medical Journal* 11 (1904–5): 56–63, p. 57; Thomas R. Marshall, *Asiatic Cholera in the Philippines Islands* (Manila: Bureau of Public Printing, 1904), p. 9; and Henry du Rest Phelan, "Sanitary service in Surigao, a Filipino town on the island of Mindanao," *Journal of the Association of Military Surgeons* 14 (1904): 1–18, p. 18.

13 Philip E. Garrison, "The prevalence and distribution of the animal parasites of man in the Philippine Islands, with a consideration of their possible influence upon the public health," *Philippine Journal of Science* 3B (1908): 191–210, pp. 205, 208.

14 Edward L. Munson, "Cholera carriers in relation to cholera control," *Philippine Journal of Science* 10B (1915): 1–9, pp. 9, 4–5.

15 Alvin J. Cox, *Thirteenth Annual Report of the Director of the Bureau of Science for the Year Ending December 31, 1914* (Manila: Bureau of Printing, 1915), p. 11.

16 Quoted in Frederick C. Chamberlin, *The Philippine Problem, 1898–1913* (Boston: Little, Brown and Co., 1913), pp. 113–14.

17 Rob Wilson, "Techno-euphoria and the discourse of the American sublime," *boundary 2* 19 (1992): 205–29, p. 208.

18 Ronald T. Takaki, "The new empire: American asceticism and the 'new navy,'" in *Iron Cages: Race and Culture in Nineteenth-Century America* (New York: Oxford University Press, 1990).

19 Warwick Anderson, "The trespass speaks: White masculinity and colonial breakdown," *American Historical Review* 102 (1997): 1343–70.

20 Bureau of Health, *The Disposal of Human Wastes*, p. 5.

21 Bureau of Health, *The Disposal of Human Wastes*, pp. 5–7, 7–9.

22 David G. Willets, "General conditions affecting the public health and diseases prevalent in the Batanes Islands, P.I.," *Philippine Journal of Science* 8B (1913): 49–58, p. 51.

23 Bureau of Health, *Proposed Sanitary Code*, Health Bulletin No. 22 (Manila: Bureau of Printing, 1920), pp. 15–17.

24 James A. LeRoy, *Philippines Life in Town and Country* (New York: G.P. Putnam's Sons, 1906), p. 24; Daniel R. Williams, *The United States and the Philippines* (New York: Doubleday, 1924), p. 125; and William B. Freer, *The Philippine Experiences of an American Teacher: A Narrative of Work and Travel in the Philippine Islands* (New York: Charles Scribner's Sons, 1906), p. 8.

25 Victor G. Heiser, *Annual Report of the Bureau of Health of the Philippine Islands, July 1, 1912–June 30, 1913* (Manila: Bureau of Printing, 1914), pp. 29–30.

26 Victor G. Heiser, Notes of 1916 Trip, Hei 2, 58.5, p. 621, Rockefeller Foundation [RF], Rockefeller Archive Center, Pocantico Hills, New York, USA [RAC].

27 Frank G. Carpenter, *Through the Philippines and Hawaii* (Garden City NY: Doubleday, Page and Co., 1925), pp. 24–5.

28 Anderson, "Excremental colonialism."

29 Henri Lefebvre, *The Production of Space* [1971], trans. Donald Nicholson-Smith (Oxford: Blackwell, 1991).

30 Friedrich Nietzsche, *The Genealogy of Morals: An Attack* [1887], trans. Francis Golffing (Garden City NY: Doubleday and Co, 1956), p. 290.

31 LeRoy, *Philippines Life*, p. 200.

32 Bruno Latour, *We Have Never Been Modern*, trans. Catherine Porter (Cambridge MA: Harvard University Press, 1993), p. 154, original emphasis.

33 On the former sense, see Clement Greenberg, "Avant-garde and kitsch," *Partisan Review* 6 (1939): 34–49.

34 Saul Friedlander, *Reflections of Nazism: An Essay on Kitsch and Death* (New York: Harper and Row, 1984).

35 Walter Benjamin, "The work of art in the age of mechanical reproduction" [1935], in *Illuminations*, ed. Hannah Arendt, trans. Harry Zohn (London: Collins/Fontana, 1973), pp. 214–18.

36 Milan Kundera, *The Unbearable Lightness of Being*, trans. Michael Henry Heim (London: Faber and Faber, 1984), p. 242.

37 Anderson, *Colonial Pathologies*. More generally, see Carl Zimring, *Clean and White: A History of Environmental Racism in the United States* (New York: New York University Press, 2015).

38 Judith Walzer Leavitt, *The Healthiest City: Milwaukee and the Politics of Health Reform* (Princeton: Princeton University Press, 1982).

39 Barbara Gutmann Rosenkrantz, *Public Health and the State: Changing Views in Massachusetts, 1842–1936* (Cambridge MA: Harvard University Press, 1972).

40 Heiser, Diary of Dr. Heiser's World Trip, 1925–26, November 22, 1925, RF, Record Group [RG] 12.1, room 104, unit 63, 5.27, RAC.

41 Heiser, Diary of Dr. Heiser's World Trip, October 1930–May 1931, January 6, 1931, RF, RG 12.1, room 104, unit 63, 5.27, RAC.

42 Heiser, Diary of Dr. Heiser's World Trip, October 1930–May 1931, January 8, 1931, RF, RG 12.1, room 104, unit 63, 5.27, RAC.

43 Yaeger to Heiser, February 11, 1930, RF, RG 1.1, series 242, box 7, folder 100, RAC.

44 Yaeger to Heiser, June 17, 1932, RF, RG 1.1, series 242, box 8, folder 101, RAC.

45 Yaeger to Heiser, June 10, 1933, RF, RG 1.1, series 242, box 7, folder 94, RAC.

46 John Ettling, *The Germ of Laziness: Rockefeller Philanthropy and Public Health in the New South* (Cambridge MA: Harvard University Press, 1981).

47 C.W. Stiles, *Hookworm Disease and the Negroes* (Hampton VA: Hampton Normal and Agricultural Institute, 1909), p. 4.

48 Charles T. Nesbitt, "The health menace of alien races," *The World's Work* 28 (1913): 74–5, pp. 74–5.

49 John Farley, *To Cast Out Disease: A History of the International Health Division of Rockefeller Foundation (1913–1951)* (Oxford: Oxford University Press, 2004); and Warwick Anderson, *The*

Cultivation of Whiteness: Science, Health, and Racial Destiny in Australia [2002] (New York: Basic Books, 2004).

50 Warwick Anderson, "Going through the motions: American public health and colonial 'mimicry,'" *American Literary History* 14 (2002): 686–719.

51 Iris Borowy, *Coming to Terms with World Health: The League of Nations Health Organization* (Frankfurt am Main: Peter Lang, 2009).

52 Edmund Wagner and J.N. Lanox, *Excreta Disposal for Rural Areas and Small Communities* (Geneva: World Health Organization, 1958); Marcos Cueto, Theodore Brown, and Elizabeth Fee, *The World Health Organization* (Cambridge: Cambridge University Press, 2019); and Iris Borowy, "Human excreta: Hazardous waste or valuable resource?" *Journal of World History* 32 (2021): 517–45

53 Elizabeth Fee, "Abel Wolman (1892–1989): Sanitary engineer of the world," *American Journal of Public Health* 101, 4 (2011): 645.

54 Homi K. Bhabha, "Of mimicry and man: The ambivalence of colonial discourse," *October* 28 (1984): 125–33.

55 Abby A. Rockefeller, "Civilization and sludge: Notes on the history of management of human excreta," *Capitalism Nature Socialism* 9, 3 (1998): 3–18, pp. 5, 11, 16, 18.

56 Aimé Césaire, *Return to my Native Land* [1939], trans. John Berger and Anne Bostock (Brooklyn NY: Archipelago Books, 2015), p. 48.

57 Karl Marx, *Capital, Volume 1* [1867], trans. Ben Fowkes (New York: Penguin, 1976); and Hayden White, "The noble savage theme as fetish," in *Tropics of Discourse: Essays in Cultural Criticism* (Baltimore: Johns Hopkins University Press, 1978), pp. 183–96.

58 William Pietz, *The Problem of the Fetish*, ed. Francesco Pellizzi, Stefanos Geroulanos, and Ben Kafka (Chicago: University of Chicago Press, 2022). On Brown, see Chapter 2.

59 Homi K. Bhabha, "The other question," *Screen* 24, 6 (1983): 18–36.

60 György Lukács, *History and Class Consciousness: Studies in Marxist Dialectics* [1923], trans. Rodney Livingstone (London: Merlin Press, 1971).

61 William Kupinse, "The Indian subject of colonial hygiene," in *Filth: Dirt, Disgust and Modern Life*, ed. William H. Cohen and Ryan Johnson (Minneapolis: University of Minnesota Press, 2005), pp. 250–76. See also Stefanos Geroulanos and Ben Kafka, "An introduction to the sheer incommensurable togetherness of the living existence of the personal self and the living otherness of the material world," in Pietz, *The Problem of the Fetish*.

Chapter 4 Powers of Ordure

1 José Rizal, *Noli me Tangere* [1887], trans. Ma. Soledad Lacson-Locsin (Honolulu: University of Hawaii Press, 1997), p. 175.

2 Rizal, *Noli me Tangere*, pp. 184, 197, 201, 202. Benedict Anderson refers to the "first Filipino" in *The Spectre of Comparisons: Nationalism, Southeast Asia, and the World* (London: Verso, 1998).

3 Mrs. Campbell Dauncey, *An Englishwoman in the Philippines* (New York: E.P. Dutton, 1906), pp. 52, 178. See Warwick Anderson, *Colonial Pathologies: American Tropical Medicine, Race, and Hygiene in the Philippines* (Durham NC: Duke University Press, 2006).

4 I had planned to discuss the excremental regimes of Nazi death camps but every trial of rendering comparable those horrors runs the unacceptable risk of normalizing or trivializing the Holocaust, implying equivalence: for more on these sites of degradation and resistance, see Primo Levi, *If This Is a Man* [1958] and *The Truce*, trans. Stuart Woolf (London: Abacus, 1991); and Terence Des Pres, *The Survivor: An Anatomy of Life in the Death Camps* (New York: Oxford University Press, 1976), chapter 3. I also refrain from analyzing events at Abu Ghraib prison in Iraq in 2003–4, where American guards ordered Arab prisoners to take shit from the toilets and smear it on their bodies.

5 Martin Pops uses the phrase in "The metamorphosis of shit,"
   Salmagundi 56 (1982): 26–61, p. 32.

6 Geoffrey Galt Harpham, *On the Grotesque: Strategies of
   Contradiction in Art and Literature* (Princeton: Princeton
   University Press, 1982).

7 Mikhail M. Bakhtin, *Rabelais and His World*, trans Hélène
   Iswolsky [1968] (Bloomington: Indiana University Press, 1984).
   See also Annabel L. Kim, "Seriously scatological," *Contemporary
   French and Francophone Studies* 27, 1 (2023): 24–32.

8 Bakhtin, *Rabelais and His World*, pp. 21, 23, 320.

9 Bakhtin, *Rabelais and His World*, pp. 26–7, 223–4.

10 François Rabelais, *Gargantua and Pantagruel* [1532–52], trans.
   Thomas Urquart and Pierre le Motteux (Mineola NY: Dover,
   2016), pp. 49, 56.

11 Rabelais, *Gargantua and Pantagruel*, pp. 56, 680, 118.

12 Bakhtin, *Rabelais and His World*, p. 175. As Annabel A. Kim
   observes, Rabelais prefigures the "unsheddable fecality" of French
   literature, which she interprets as a form of "fecal universalism,"
   a sign of radical equality (*Cacaphonies: The Excremental Canon
   of French Literature* [Minneapolis: University of Minnesota
   Press, 2022], pp. 2, 23). Kim focuses on later figures such as
   Louis-Ferdinand Céline, Samuel Beckett, and Jean Genet.

13 J. Middleton Murry used the term "excremental vision" in
   Jonathan Swift: A Critical Biography (London: Jonathan Cape,
   1954). See Everett Zimmerman, "Swift's scatological poetry:
   A praise of folly," *Modern Language Quarterly* 48, 2 (1987):
   124–44; and Ashraf H.A. Rushdy, "A new emetics of interpretation: Swift, his critics and the alimentary canal," *Mosaic* 24, 3/4
   (1991): 1–32.

14 Norman O. Brown, *Life Against Death: The Psychoanalytic
   Meaning of History* [1959], 2nd edition (Middletown CT:
   Wesleyan University Press, 1985), p. 200. See also Peter J. Smith,
   *Between Two Stools: Scatology and its Representations in English
   Literature, Chaucer to Swift* (Manchester: Manchester University
   Press, 2015).

15 Jonathan Swift, *Gulliver's Travels into Several Remote Nations of the World* [1726] (Minneapolis: Lerner Publishing, 2014), pp. 32, 220.

16 Swift, *Gulliver's Travels*, pp. 295, 305, 312.

17 Stephen Greenblatt, "Filthy rites," *Daedalus* 3, 3 (1982): 1–16, p. 4.

18 Peter Stallybrass and Allon White, *The Politics and Poetics of Transgression* (Ithaca: Cornell University Press, 1986), p. 6.

19 Stallybrass and White, *Politics and Poetics of Transgression*, p. 18.

20 Roland Barthes, *Sade, Fourier, Loyola* (Paris: Seuil, 1971), p. 140.

21 James Joyce, *Ulysses* [1922] (New York: Random House, 1961). Kelly Anspaugh uses "cloacal imperialism" in "Ulysses upon Ajax? Joyce, Harrington, and the question of 'cloacal imperialism,'" *South Atlantic Review* 60, 2 (1995): 11–29. See also Kelly Anspaugh, "Powers of ordure: James Joyce and the excremental vision(s)," *Mosaic* 27, 1 (1994): 73–10.

22 Joyce, *Ulysses*, pp. 131, 133, original emphasis.

23 Achille Mbembe, "The banality of power and the aesthetics of vulgarity in the postcolony," trans. Janet Roitman, *Public Culture* 4, 2 (1992): 1–30, pp. 1, 4.

24 Mbembe, "The banality of power," pp. 5, 12, 16.

25 Achille Mbembe, *On the Postcolony* (Berkeley: University of California Press, 2001), pp. 110, 123, 126.

26 Judith Butler, "Mbembe's extravagant power," *Public Culture* 5, 1 (1992): 67–74, pp. 68, 73.

27 Joshua D. Esty, "Excremental postcolonialism," *Contemporary Literature* 40, 1 (1999): 22–59, pp. 31, 30. See also Warwick Anderson, "Crap on the map, or postcolonial waste," *Postcolonial Studies* 13, 2 (2010): 169–78.

28 Wole Soyinka, *The Interpreters* [1965] (Portsmouth: Heinemann, 1970), pp. 72, 108.

29 Ayi Kwei Armah, *The Beautiful Ones Are Not Yet Born* (Portsmouth: Heinemann, 1969), p. 125.

30 Edward W. Said, "Yeats and decolonization," in *Nationalism, Colonialism, and Literature* (Minneapolis: University of Minnesota Press, 1990), pp. 69–95, p. 85.

31 John G. Bourke, *The Scatologic Rites of All Nations. A Dissertation upon the Employment of Excrementitious Remedial Agents in Religion, Therapeutics, Divination, Witchcraft, Love-Philters, etc., in All Parts of the Globe* (Washington DC: W.H. Lowdermilk, 1891).

32 Sigmund Freud, "Foreword" [1913], in *The Portable Scatalog: Excerpts from Scatalogic Rites of All Nations by John G. Bourke*, ed. Louis P. Kaplan (New York: William Morrow, 1994), pp. 5–9.

33 Bronislaw Malinowski, *Argonauts of the Western Pacific: An Account of Native Enterprise and Adventure in the Archipelagoes of Melanesian New Guinea* (London: Routledge and Kegan Paul, 1922).

34 Sjaak van der Geest, "Not knowing about defecation," in *On Knowing and Not Knowing in the Anthropology of Medicine*, ed. Roland Littlewood (New York: Routledge, 2007), pp. 75–86, p. 78.

35 Bronislaw Malinowski, *Sex and Repression in Savage Society* (London: Kegan Paul Trench, Trubner, 1927). On the expedition more generally, see Warwick Anderson, "Hermannsburg, 1929: Turning Aboriginal 'primitives' into modern psychological subjects," *Journal of the History of the Behavioral Sciences* 50 (2014): 127–47.

36 Weston La Barre, "Géza Róheim, 1891–1953: Psychoanalysis and anthropology," in *Psychoanalytic Pioneers*, ed. Franz Alexander, Samuel Eisenstein, and Martin Grotjahn (New York: Basic Books, 1966), pp. 272–81, pp. 272, 275.

37 Géza Róheim, "Freud and cultural anthropology," *Psychoanalytic Quarterly* 9 (1940): 246–55, p. 255.

38 Géza Róheim, *The Eternal Ones of the Dream: A Psychoanalytic Interpretation of Australian Myth and Ritual* (New York: International Universities Press, 1945), and "Dream analysis and fieldwork in anthropology," in *Man and his Culture:*

Psychoanalytic Anthropology after Totem and Taboo, ed. Werner Muensterberger (New York: Taplinger Publishing, 1970), pp. 139–75.

39 Géza Róheim, "The psychoanalysis of primitive cultural types," *International Journal of Psychoanalysis* 13 (1932): 1–222, p. 85.

40 Róheim interview, 1952, folder 6, box 121, Sigmund Freud Papers, Manuscripts Division, Library of Congress, Washington DC.

41 Sjaak van der Geest, "Akan shit: Getting rid of dirt in Ghana," *Anthropology Today* 14, 3 (1998): 8–12, and "The night soil collector: Bucket latrines in Ghana," *Postcolonial Studies* 5, 2 (2002): 197–206.

42 Van der Geest, "The night soil collector," pp. 199, 220.

43 Brenda Chalfin, "Public things, excremental politics, and the infrastructure of bare life in Ghana's city of Tema," *American Ethnologist* 41 (2014): 92–109, pp. 93, 105. See also her "'Wastelandia': Infrastructure and the commonwealth of waste in urban Ghana," *Ethnos* 82 (2016): 648–71.

44 Arjun Appadurai, "Deep democracy: Urban governmentality and the horizons of politics," *Environment and Urbanization* 13, 2 (2001): 23–43, pp. 23, 37, 36, 37.

45 Tulasi Srinivas, "Flush with success: Bathing, defecation, worship and social change in South India," *Space and Culture* 5, 4 (2002): 368–86, pp. 371, 370.

46 Diane Coffey, Aashish Gupta, Payal Hathi, Dean Spears et al., "Understanding open defecation in rural India: Untouchability, pollution, and latrine pits," *Economic and Political Weekly* 52, 1 (2017): 59–66, pp. 59, 61. See also Diane Coffey, Aashish Gupta, Payal Hathi, Nidhi Khurana et al., "Revealed preference for open defecation," *Economic and Political Weekly* 49, 38 (2014): 43.

47 Assa Doron and Robin Jeffrey, "Open defecation in India," *Economic and Political Weekly* 49, 49 (2014): 72–8, p. 73. See also Assa Doron and Ira Raja, "The cultural politics of shit: Class, gender and public spaces in India," *Postcolonial Studies* 18, 2

(2015): 189–207; and Assa Doron and Robin Jeffrey, *Waste of a Nation: Garbage and Growth in India* (Cambridge MA: Harvard University Press, 2018).

48 Sarah Jewitt, "Geographies of shit: Spatial and temporal variations in attitudes towards human waste," *Progress in Human Geography* 35, 5 (2011): 608–26, p. 615.

49 For example, Sophia Stamatopoulou-Robbins, *Waste Siege: The Life of Infrastructure in Palestine* (Stanford: Stanford University Press, 2019); Indrawan Prabaharyaka, "Shit, shit, every where (or: notes on the difficulties of classifying shits)," *International Development Planning Review* 42, 3 (2020): 295–313; Dean Chahim, "The logistics of waste: Engineering, capital accumulation, and the growth of Mexico City," *Antipode* (2022): https://doi.org/10.1111/anti.12864; and Kathrine Eitel, "Reshuffling responsibility: Waste, environmental justice and urban citizenship in Cambodia," *Worldwide Waste: Journal of Interdisciplinary Studies* 5, 1 (2022): 1–13. For how fears of making feces accessible to sorcerers shape excretory cultures in the eastern highlands of New Guinea see Shirley Lindenbaum, *Kuru Sorcery: Disease and Danger in the New Guinea Highlands* (Palo Alto: Mayfield, 1979) and Warwick Anderson, *The Collectors of Lost Souls: Turning Kuru Scientists into Whitemen* (Baltimore: Johns Hopkins University Press, 2008).

50 Josh Dawsey, "Trump derides protections for immigrants from 'shithole' countries," *Washington Post* (2018): https://www.washingtonpost.com/politics/trump-attacks-protections-for-immigrants-from-shithole-countries-in-oval-office-meeting/2018/01/11/bfc0725c-f711–11e7–91af–31ac729add94_story.html.

51 Arturo Escobar, *Encountering Development: The Making and Unmaking of the Third World* (Princeton: Princeton University Press, 1995); and Akhil Gupta, *Postcolonial Developments: Agriculture in the Making of Modern India* (Durham NC: Duke University Press, 1998).

52 Paul Theroux, *The Great Railway Bazaar: By Train Through Asia* (Boston: Houghton Mifflin, 1975), p. 157.

53 Simon Watney, "Missionary positions: AIDS, 'Africa,' and race," in *Out There: Marginalization and Contemporary Cultures*, ed. Russell Ferguson, Martha Gever, Trinh T. Minh-ha, and Cornel West (Cambridge MA: MIT Press, 1990), pp. 89–103; and Cindy Patton, "From nation to family: Containing 'African AIDS,'" in *Nationalisms and Sexualities*, ed. Andrew Parker et al. (New York: Routledge, 1992), pp. 218–34.

54 Jean Comaroff, "Beyond bare life: AIDS, (bio)politics, and the neoliberal order," *Public Culture* 19 (2007): 197–219, pp. 197, 198, 211.

55 Linda S. Mitteness and Judith C. Barker, "Stigmatizing a 'normal' condition: Urinary incontinence in late life," *Medical Anthropology Quarterly* 9 (1995): 188–210; and Els Van Dongen, "It isn't something to yodel about, but it exists! Faeces, nurses, social relations and status within a mental hospital," *Aging and Mental Health* 5, 3 (2001): 205–15.

56 Julia Lawton, "Contemporary hospice care: The sequestration of the unbounded body and 'dirty dying,'" *Sociology of Health and Illness* 20, 2 (1998): 121–43.

57 Sjaak van der Geest and Shahaduz Zaman, "'Look under the sheets!': Fighting with the senses in relation to defecation and bodily care in hospitals and care institutions," *Medical Humanities* 47 (2021): 103–11.

58 Joiada Verrips, "Excremental art: Small wonder in a world full of shit," *Journal of Extreme Anthropology* 1, 1 (2017): 19–46, p. 32.

59 Hal Foster, "The 'primitive' unconscious of modern art," *October* 34 (1985): 45–70.

60 Sigmund Freud, "Character and anal eroticism" [1908], in *The Standard Edition of the Complete Psychological Works of Sigmund Freud*, trans. James Strachey, volume 9 (London: Hogarth Press, 1952), pp. 167–75; and Ernest Jones, "Anal-erotic character traits," *Journal of Abnormal Psychology* 13, 5 (1908): 261–84.

61 Hal Foster, "'Primitive' scenes," *Critical Inquiry* 20 (1993): 69–102, pp. 72, 93.

62 Gerald Silk, "Myths and meanings in Manzoni's *Merda d'artista*," *Art Journal* 52, 3 (1993): 65–75, p. 65. Among the many white blokes into shit art are Piero Manzoni, Mike Kelley, Bob Flanagan, Robert Mapplethorpe, Gilbert and George, Paul McCarthy, and Wim Delvoye – though exceptions include Andres Serrano, Chris Ofili, Terence Koh, Kiki Smith, and Karen Finley. If space permitted, I would discuss the minting of a non-fungible token (NFT) from human fecal matter by "The Most Famous Artist," Matty Mo (Monahan), based in Los Angeles, in 2021. Mo auctioned the token to raise money for research into possible links between the human gut microbiome and autism.

63 John Miller, "Excremental value," *Tate Etc.* 10 (2007): 40–3.

64 Wim Delvoye, *Cloaca* (New York: New Museum of Contemporary Art, 2001).

65 Quoted in Robert Enright, "Vim and vigour: An interview with Wim Delvoye," *Bordercrossings* (2005): https://bordercrossingsmag.com/article/vim-and-vigour-an-interview-with-wim-delvoye.

66 Quoted in Anon., "I've always tried to make art that could be understood by everybody," *Apollo* (2017): https://www.thefree library.com/%27I%27ve+always+tried+to+make+art+that+coul d+be+understood+by+everybody%27.-a0497797670.

67 Quoted in Lynn Yaeger, "Andres Serrano's 'shit' show," *Village Voice* (2008): https://www.villagevoice.com/2008/08/27/andres -serranos-shit-show.

68 Quoted in Elizabeth Bumiller, "Civilization, sanitation, and the mayor," *New York Times* (13 October 1999): B5.

69 Foster, "'Primitive' scenes," p. 103, original emphasis.

70 Judith Keegan Gardiner, "*South Park*, blue men, anality, and market masculinity," *Men and Masculinities* 2, 3 (2000): 251–71, p. 252.

71 David Larsen, "*South Park*'s solar anus, or, Rabelais returns: Cultures of consumption and the contemporary aesthetic of obscenity," *Theory, Culture and Society* 18, 4 (2001): 65–82.

72 Jonathan Swift, *A Tale of a Tub* [1704] (New York: Columbia University Press, 1930), pp. 25, 25–6.

73 Guy Debord, *The Society of the Spectacle* [1967], trans. Donald Nicolson-Smith (New York: Zone, 1994).

74 James C. Scott, *Weapons of the Weak: Everyday Forms of Peasant Resistance* (New Haven: Yale University Press, 1985).

75 Allen Feldman, *Formations of Violence: The Narrative of the Body and Political Terror in Northern Ireland* (Chicago: University of Chicago Press, 1991).

76 Begoña Aretxaga, "Dirty protest: Symbolic overdeteration and gender in Northern Ireland ethnic violence," *Ethos* 23 (1995): 123–48, pp. 129, 135, 136.

77 Georges Bataille, *Visions of Excess: Selected Writings, 1927–1939*, trans. Allan Stoekl, Carl R. Lovitt, and Donald M. Leslie, Jr. (Minneapolis: University of Minnesota Press, 1985), and *Divine Filth: Lost Scatology and Erotica*, trans. Mark Spitzer (London: Creation, 2004).

78 Vicente Rafael, *The Sovereign Trickster: Death and Laughter in the Age of Duterte* (Durham NC: Duke University Press, 2022), p. 102.

Chapter 5 Gut Feelings and Dark Continents

1 Ed Yong, *I Contain Multitudes: The Microbes Within Us and a Grander View of Life* (London: Harper Collins, 2016).

2 Stefan Helmreich, "Trees and seas of information: Alien kinship and the biopolitics of gene transfer in marine biology and biotechnology," *American Ethnologist* 30 (2003): 30–58.

3 Heather Paxson, "Post-Pasteurian cultures: The microbiopolitics of raw-milk cheese in the United States," *Cultural Anthropology* 23 (2008): 15–47.

4 Heather Paxson and Stefan Helmreich, "The perils and promises of microbial abundance: Novel natures and model ecosystems, from artisanal cheese to alien seas," *Social Studies of Science* 44, 2 (2014): 165–93, pp. 165, 167.

5 Davina Höll and Leonie N. Bossert, "Introducing the microbiome: Interdisciplinary perspectives," *Endeavour* 46 (2022): e100817.

6 Joshua Lederberg and Alexa T. McCray, "'Ome sweet 'omics: A genealogical treasury of words," *Scientist* 15 (2001): 80.

7 Georg Simmel, *The Philosophy of Money* [1900], ed. David Frisby, trans. Tom Bottomore and David Frisby (London: Routledge, 2004), pp. 69, 444, 87.

8 Robert P. Hudson, "Theory and therapy: Ptosis, stasis, and autointoxication," *Bulletin of the History of Medicine* 63 (1989): 392–413; Micaela Sullivan-Fowler, "Doubtful theories, drastic therapies: Autointoxication and faddism in the late nineteenth and early twentieth centuries," *Journal of the History of Medicine and Allied Sciences* 50 (1995): 364–90; James C. Whorton, *Inner Hygiene: Constipation and the Pursuit of Health in Modern Society* (Oxford: Oxford University Press, 2000); Ian Miller, *A Modern History of the Stomach: Gastric Illness, Medicine, and British Society, 1800–1950* (London: Pickering and Chatto, 2011); and Manon Mathias, "Autointoxication and the historical precursors of the microbiome-gut-brain axis," *Microbial Ecology in Health and Disease* 29, 2 (2018): https://doi.org/10.1080/1651 2235.2018.1548249.

9 Émile Gautier, "Thérapeutique: pour ne pas vieillir trop vite," *Annales politiques et littéraires* (7 July 1907): 17–18, pp. 17, 18.

10 Alfred I. Tauber and Leon Chernyak, *Metchnikoff and the Origins of Immunology* (Oxford: Oxford University Press, 1991).

11 Élie Metchnikoff, *The Nature of Man: Studies in Optimistic Philosophy*, trans. P. Chalmers Mitchell (London: Heinemann, 1903), and *The Prolongation of Life: Optimistic Studies*, trans. P. Chalmers Mitchell (New York: G.P. Putnam's Sons, 1907).

12 Charles Bouchard, *Leçons sur les auto-intoxications dans les maladies* (Paris: Librairie F. Savy, 1887).

13 Erwin H. Ackerknecht, "Broussais or a forgotten medical revolution," *Bulletin of the History of Medicine* 27 (1953): 320–43.

14 Kirill Rossiianov, "Taming the primitive: Élie Metchnikov and his discovery of immune cells," *Osiris* 23, 1 (2008): 213–29.

15 Metchnikoff, *The Nature of Man*, p. 239.

16 H.G. Wells, "The philosophy of a biologist," *The Speaker* (31 October 1903): 112.

17 H.G. Wells, *The History of Mr Polly* [1910] (London: Penguin, 2005). See also Paul Vlitos, "Unseen battles: H.G. Wells and auto-intoxication theory," *The Wellsian: The Journal of the H.G. Wells Society* 36 (2013): 25–38.

18 W. Arbuthnot Lane, "An address on chronic intestinal stasis," *British Medical Journal* (June 12, 1909): 1408–11, p. 1410. See also W. Arbuthnot Lane, *The Operative Treatment of Chronic Constipation* (London: Nisbet, 1909).

19 J.L. Smith, "Sir Arbuthnot Lane, chronic intestinal stasis, and autointoxication," *Annals of Internal Medicine* 96 (1982): 365–9.

20 John Harvey Kellogg, *Autointoxication or Intestinal Toxemia* (Battle Creek: Modern Medicine Publishing Co., 1919).

21 John Harvey Kellogg, *Itinerary of a Breakfast: A Popular Account of the Travels of a Breakfast Through the Feed Tube...* (New York: Funk and Wagnalls, 1920), pp. 4, 63, 23, 91, 159, 93.

22 Walter C. Alvarez, "Origin of the so-called autointoxication symptoms," *JAMA* 72, 1 (1919): 8–13, pp. 11, 12, 8.

23 Arthur N. Donaldson, "Relation of constipation to intestinal intoxication," *JAMA* 78, 12 (1922): 884–8, pp. 884, 885, 886.

24 Alexander R.P. Walker, "Nutritionally related disorders/diseases in Africans: Highlights in half a century of research, with special reference to unexpected phenomena," in *Dietary Fiber in Health and Disease*, ed. David Kritchevsky and Charles Bonfield (New York: Plenum Press, 1997), pp. 1–14.

25 Denis K. Burkitt, Alexander R.P. Walker, and Neil S. Painter, "Effect of dietary fibre on stools and transit-times, and its role in the causation of disease," *The Lancet* 300, 7792 (1972): 1408–11.

26 Sebastián Gil-Riaño and Sarah E. Tracy, "Developing constipation: Dietary fiber, Western disease, and industrial carbohydrates," *Global Food History* 2, 2 (2016): 179–209, p. 198.

27 Charles E. Rosenberg, "Pathologies of progress: The idea of civilization as risk," *Bulletin of the History of Medicine* 72 (1998): 714–30.

28 Sonja van Wichelen tells me that a PubMed search reveals sev-
    enty-four publications that mentioned the "human microbiome"
    in 2001, and 21,648 in 2021. It would seem the twenty-first cen-
    tury is becoming the age of the microbiome – and of excremental
    inquiry more generally.

29 Julian Davies, "In a map for human life, count the microbes, too,"
    Science 291, 5512 (2001): 2316.

30 David A. Relman and Stanley Falkow, "The meaning and impact
    of the human genome sequence for microbiology," *Trends in
    Microbiology* 9, 5 (2001): 206–8.

31 International Human Genome Sequencing Consortium, "Initial
    sequencing and analysis of the human genome," *Nature* 409
    (2001): 860–921; and Roswitha Löwer, Johannes Löwer, and
    Reinhard Kurth, "The viruses in all of us: Characteristics and bio-
    logical significance of human endogenous retroviral sequences,"
    Proceedings of the National Academy of Sciences (USA) 93, 11
    (1996): 5177–84.

32 Relman and Falkow, "The meaning and impact of the human
    genome sequence for microbiology," pp. 206, 208.

33 Jo Handelsman, Michelle R. Rondon, Sean F. Brady, Jon Clardy,
    and Robert M. Goodman, "Molecular biological access to the
    chemistry of unknown soil microbes: A new frontier for natu-
    ral products," *Chemistry and Biology* 5, 10 (1998): R245–9; and
    Duccio Medini, Davide Serruto, Julian Parkhill, David A. Relman
    et al., "Microbiology in the post-genomic era," *Nature Reviews
    Microbiology* 6 (2008): 419–30.

34 Peter J. Turnbaugh, Ruth E. Ley, Micah Hamady, Claire M.
    Fraser-Liggett et al., "The Human Microbiome Project," *Nature*
    449 (2007): 804–10.

35 Jane Peterson, Susan Garges, Maria Giovanni, Pamela McInnes
    et al., "The NIH Human Microbiome Project," *Genome Research*
    19 (2009): 2317–23, pp. 2317, 2319.

36 Steven Epstein, *Inclusion: The Politics of Difference in
    Medical Research* (Chicago: University of Chicago Press,
    2007).

37 Barbara A. Methé, Karen E. Nelson, Mihai Pop, Heather H. Creasy et al., "A framework for human microbiome research," *Nature* 486, 7402 (2012): 215–21, p. 221.

38 P. Forsythe, W. Kunze, and J. Bienenstock, "Moody microbes and fecal phrenology: What do we know about the microbiota-gut-brain axis?" *BMC Medicine* 14, 58 (2016): https://doi.org/10.11 86/s12916–016–0604–8.

39 Maureen A. O'Malley and Derek J. Skillings, "Methodological strategies in microbiome research and their explanatory implications," *Perspectives on Science* 26, 2 (2018): 239–65.

40 Joanna Radin and Emma Kowal, eds., *Cryopolitics: Frozen Life in a Melting World* (Cambridge MA: MIT Press, 2017).

41 Michael Pollan, "Some of my best friends are germs," *New York Times* (2013): https://www.nytimes.com/2013/05/19/magazine /say-hello-to-the–100-trillion-bacteria-that-make-up-your-microbiome.html.

42 Lynn Margulis, ed. *Symbiosis as a Source of Evolutionary Innovation: Speciation and Morphogenesis* (Cambridge MA: MIT Press, 1991). See Scott F. Gilbert, Jan Sapp, and Alfred I. Tauber, "A symbiotic view of life: We have never been individuals," *Quarterly Review of Biology* 87, 4 (2012): 325–41.

43 Stefan Helmreich, "*Homo microbis*: The human microbiome, figural, literal, political," *Thresholds* 42 (2014): 52–9, p. 56.

44 Tobias Rees, Thomas Bosch, and Angela E. Douglas, "How the microbiome challenges our concept of self," *PLoS Biology* 16, 2 (2018): e2005358.

45 Maureen A. O'Malley, *Philosophy of Microbiology* (Cambridge: Cambridge University Press, 2014); and Emily C. Parke, Brett Calcott, and Maureen A. O'Malley, "A cautionary note for claims about the microbiome's impact on the 'self,'" *PLoS Biology* 16, 9 (2018): e2006654.

46 Katarzyna B. Hooks and Maureen A. O'Malley, "Dysbiosis and its discontents," *mBio* 8, 5 (2017): e01492–17.

47 Anahad O'Connor, "The best foods to feed your gut microbiome," *Washington Post* (2022): https://www.washingtonpost.com/wellness/2022/09/20/gut-health-microbiome-best-foods.

48 Yong, *I Contain Multitudes*.

49 Alex Nading, "Evidentiary symbiosis: On paraethnography in human-microbe relations," *Science as Culture* 25, 4 (2016): 560–81.

50 José Clemente, E.C. Pehrsson, M.J. Blaser, K. Sandhu et al., "The microbiome of uncontacted Amerindians," *Scientific Advances* 3, 1 (2015): e1500183.

51 Richard J. Abdill, Elizabeth M. Adamowicz, and Ran Blekhman, "Public human microbiome data are dominated by highly developed countries," *PLoS Biology* 20, 2 (2022): e3001536.

52 Amber Benezra, "Race in the microbiome," *Science, Technology and Human Values* 45, 5 (2020): 877–90, p. 883. See also Hiʻilei Julia Hobart and Stephanie Maroney, "On racial constitutions and digestive therapeutics," *Food, Culture, and Society* 22, 5 (2019): 576–94.

53 Quoted in Emily Eakin, "The excrement experiment: Treating disease with fecal transplants," *The New Yorker* (2014): https://www.newyorker.com/magazine/2014/12/01/excrement-experiment.

54 Erica D. Sonnenburg and Justin L. Sonnenburg, "The ancestral and industrialized gut microbiota and implications for human health," *Nature Reviews Microbiology* 17 (2019): 383–90; and Samuel A. Smits, Jeff Leach, Erica D. Sonnenburg, Carlos G. Gonzalez et al., "Seasonal cycling in the gut microbiome of the Hadza, hunter-gatherers of Tanzania," *Science* 357, 6353 (2017): 802–6.

55 That they spoke Swahili suggests they may not have been so "uncontaminated." In any case, as I recall from my days as a medical student at the Kenyatta Hospital in Nairobi, the common Swahili word for shit is "shit," with "mavi" apparently meaning animal dung.

56 Quoted in Eakin, "The excrement experiment."

57 Tim Spector and Jeff Leach, "I spent three days as a hunter-gatherer to see if it would improve my gut health," *The Conversation* (2017): https://theconversation.com/i-spent-three-days-as-a-hunter-gatherer-to-see-if-it-would-improve-my-gut-health–7 8773.

58 Gina Kolata, "You're missing microbes. But is 'rewilding' the way to get them back?" *New York Times* (2021): https://www.nytimes.com/2021/07/19/health/human-microbiome-hadza-rewilding.html. See also Lief Reigstad, "They accused a man of sexual assault in a small West Texas town. That was only the beginning," *Texas Monthly* (2021): https://www.texasmonthly.com/news-politics/they-accused-a-man-of-sexual-assault-in-a-small-west-texas-town-that-was-only-the-beginning.

59 David P. Strachan, "Hay fever, hygiene, and household size," *British Medical Journal* 299, 6710 (1989): 1259–60. See also Warwick Anderson and Ian R. Mackay, *Intolerant Bodies: A Short History of Autoimmunity* (Baltimore: Johns Hopkins University Press, 2014).

60 Mark S. Wilson and Rick M. Maizels, "Regulation of allergy and autoimmunity in helminth infection," *Clinical Reviews in Allergy and Immunology* 26, 1 (2014): 35–50; and Graham A.W. Rook, "Review series on helminths, immune modulation, and the hygiene hypothesis: The broader implications of the hygiene hypothesis," *Immunology* 126, 1 (2009): 3–11.

61 Jamie Lorimer, "Gut buddies: Multispecies studies and the microbiome," *Environmental Humanities* 8, 1 (2016): 57–76, p. 57.

62 Tim Adams, "Gut instinct: The miracle of the parasitic hookworm," *The Observer* (2010); https://www.theguardian.com/lifeandstyle/2010/may/23/parasitic-hookworm-jasper-lawrence-tim-adams.

63 Moises Velasquez-Manoff, "An epidemic of absence: Destroying bugs in our bodies can be dangerous to our health," *The Daily Beast* (2012): https://www.thedailybeast.com/an-epidemic-of-absence-destroying-the-bugs-in-our-bodies-can-be-dangerous-to-our-health.

64 William Parker, "We need worms," *Aeon* (2019): https://aeon.co /essays/gut-worms-were-once-a-cause-of-disease-now-they-are -a-cure. See also William Parker and Jeff Ollerton, "Evolutionary biology and anthropology suggest biome reconstitution as a necessary approach toward dealing with immune disorders," *Evolution, Medicine, and Public Health* 1 (2013): 89–103.

65 Matthew J. Wolf-Meyer, "Normal, regular, and standard: Scaling the body through fecal microbial transplants," *Medical Anthropology Quarterly* 31, 3 (2016): 297–314.

66 Ben Eiseman, W. Silen, G.S. Bascom, and A.J. Kauvar, "Fecal enema as an adjunct in the treatment of pseudomembranous enterocolitis," *Surgery* 44, 5 (1958): 854–9, p. 859.

67 Brooke C. Wilson, Tommi Vatanen, Wayne Si Cutfield, and Justin M. O'Sullivan, "The super-donor phenomenon in fecal microbiota transplantation," *Frontiers in Cellular and Infection Microbiology* 9 (2019): https://doi.com/10.3389/fcimb.2019 .00002.

68 Eakin, "The excrement experiment."

69 Justin Chen, Amanda Zaman, Bharat Ramakrishna, Scott W. Olesen et al., "Stool banking for fecal microbiota transplantation: Methods and operations at a large stool bank," *Frontiers in Cellular and Infection Microbiology* 11 (2021): e662949.

70 Eakin, "The excrement experiment."

71 The website https://thepowerofpoop.com was accessed freely in March 2022 but at the time of writing is unavailable without subscription. The use of the American "poop" suggests an ambition to appeal to global markets.

72 Lawrence Cohen, "Operability, bioavailability, and exception," in *Global Assemblages: Technology, Politics, and Ethics as Anthropological Problems*, ed. Aihwa Ong and Stephen J. Collier (Oxford: Blackwell, 2005), pp. 79–90.

Conclusion

1 Elias Canetti, *Crowds and Power* [1960], trans. Carol Stewart (New York: Farrar, Straus, and Giroux, 1984), p. 265.

2 Canetti, *Crowds and Power*, p. 266.

3 W.H. Auden, "The geography of the house," in *About the House* (New York: Random House, 1965), p. 17.

4 Guy Debord, *The Society of the Spectacle* [1967], trans. Ken Knabb (London: Rebel Press, 2006), p. 13.

5 Jean Baudrillard, "Consumer society," in *Selected Writings*, ed. Mark Poster, trans. Jacques Mourrais (Stanford: Stanford University Press, 1988), p. 19.

6 Jonathan Swift, *Meditation Upon a Broomstick* (London: E. Curll, 1714), p. 4.

Acknowledgments

1 Erving Goffman, *Stigma: Notes on the Management of Spoiled Identity* (New York: Touchstone, 1963).

Index